A Window on Intelligence
Dennis Hackethal

A Window on Intelligence

The Philosophy of People, Software, and Evolution – and Its Implications

Dennis Hackethal

ISBN 978-1-7346961-3-4 (paperback)
ISBN 978-1-7346961-4-1 (.epub ebook)
ISBN 978-1-7346961-5-8 (.mobi ebook)

First edition March 2020
Second edition June 2020

Dennis Hackethal, Cupertino 2020
Illustrations depicting communications travel in chapter 10 by Carla Lagemann

CONTENTS

	ACKNOWLEDGMENTS	*i*
	INTRODUCTION	*ii*
1	A Brief Account of the Origin of Computer Programs	1
2	How Do We Know?	8
3	Knowledge in Computers – A Functional Approach	30
4	Universality	51
5	Creativity	64
6	All People Are Software Engineers	142
7	Current Intelligence Research	149
8	Ethics and Dangers of Intelligence Research	192
9	A Programmer's View of Psychotherapy	210
10	Living Among the Stars	237
	CONCLUSION	273
	REFERENCES	277
	ENDNOTES	284

Acknowledgments

I thank my friends and colleagues Thatchaphol Saranurak, Bruce Nielson, and Ella Hoeppner for reading parts of this book before its publication, providing criticism, and suggesting improvements, and David Deutsch, whose books were the main inspiration for this book, for answering my many questions. Thank you to my sister and illustrator Carla Lagemann for making some of the illustrations in this book. Last but not least, I want to thank my high-school teacher Frau Dr. Kant for recommending Popper to me.

INTRODUCTION

The story of how the human mind evolved is almost too good to be true. Contrary to most evolutionary changes, this change was not gradual. It was a jump from a cold, automated machine to a sentient, creative mind, and was, therefore, a momentous change in the history of the universe. For the first time, a part of the world could understand itself – through people. They were the first to experience what they saw whenever they opened their eyes. In front of them was a bleak, hostile world. Yet the very genetic mutation that allowed them to become aware of their surroundings also enabled them to *explain* the world, and, therefore, to control and improve it.

From this crucial moment in our evolutionary history onward, human progress chugged along. Then, the enlightenment sped it up significantly (Deutsch 2012a, 29). People have enjoyed rapid progress and technological innovations ever since. And even though hundreds of thousands of years have passed since this seemingly miraculous mutation occurred, we can still trace today's innovations and abilities back to it. It is this ability to be intelligent and create new knowledge that is also known as *creativity* (Deutsch 2012a, 30).

One of the most impactful innovations is the computer. Ever since this invention, people have tried to write intelligent programs. When the late Alan Turing, the father of modern computer science, wrote about "AI" or "artificial intelligence," the term meant something different from today. He used it to describe a program that has human intelligence. Since then, the use of the term has changed to refer to specific types of software with a certain level of sophistication, particularly those that do not need to be "explicitly" programmed, such as self-driving cars, master chess players, and text-prediction systems. Thus the term "AGI" – "artificial *general* intelligence" – had to be introduced to refer to the original concept of human intelligence again (Deutsch 2012b).

Some people have put forward other terms such as "strong AI" (Searle 2009) and "real intelligence" (Hawkins and Blakeslee, 2004), but I use "AGI" as a stand-in for all of them because, in reality, they are all equivalent. Likewise, when I say "artificial," I do not mean that it is not genuine intelligence – it *is* intelligence, only running on a computer. Hardware differences aside, it is still the real deal.

Current AI research is *narrow*. AI can only work for specific applications and cannot have a mind of its own. It cannot learn anything its programmers have not designed it to "learn." For these reasons, I shall refer to this kind of AI as "narrow AI," whereas I will refer to human intelligence as "AGI." It refers to an instance of the creative program on a computer other than the human brain. It is this creative program that makes people, people. The project of AGI attempts to replicate it and all other cognitive abilities of people that creativity enables, such as consciousness and free will.

To develop an AGI has been the goal for decades – and it would be a great achievement. While we have made a substantial amount of progress in *narrow*-AI research, we have made none toward AGI (Deutsch 2012a, 152). Why is that? It has been the toughest problem to crack since those days Alan Turing wrote his "Computing Machinery and Intelligence" and has attracted some of the brightest minds in computer science. Yet they have all failed to build AGI. It is not for lack of trying or thinking. Following the physicist, philosopher, and father of quantum computing David Deutsch, in this book I argue that it is because they have so far had the wrong *philosophy*.

There is a widespread prejudice that anything to do with philosophy is hand-wavy, pointless navel-gazing. It is this prejudice that has prevented progress in AGI research. I worry that many software engineers have it because most people in most fields do. While there is indeed much bad philosophy out there, those who fear that philosophy is generally a waste of time, fear not: there exist real philosophical problems that require solving. How to build an AGI is one of them, and it is soluble.

Philosophy is crucial: it tells us how to think. It determines our every endeavor and our chances for success. Everyone has a philosophy because each of us has a way of going about solving problems. That way *is* one's philosophy.

Software engineers, in particular, should not dismiss philosophy for the simple reason that they routinely contribute to the philosophical field of epistemology without realizing it (Temple 2010): by discovering rules governing the improvement of the structure of programs. Our understanding of what makes some implementations of a program better than others has dramatically improved since David Deutsch recently made important epistemological discoveries (2012a), which I will go over. This book is an application of some of his discoveries to software engineering, and, in turn, an application of software-engineering principles to progress in all domains. For, the search for good explanations is what drives human progress (Deutsch 2012a, vii), and it can be stated as a software-engineering principle.

Progress in software engineering depends on a single activity: writing *good* implementations in response to problems. How to write good implementations cannot be known without a good epistemology. It is a programmer's philosophy that tells him how to write his programs. All software engineers are philosophers, and all people are software engineers, whether they realize this or not.

Epistemology is the study of knowledge. It provides answers to questions such as: how is knowledge created? How does it grow? These questions are crucial because, by definition, an AGI is a program that can create knowledge of any kind: it is a *universal explainer* (Deutsch 2012a, 146). It can explain anything people can; indeed, it *is* a person (Deutsch 2012a, 157). And the study of epistemology *is* the study of AGI. Therefore, any intelligence research not directed toward epistemology is futile (Deutsch 2012b).

We know of only *one* approach to building AGI. Despite appearances, the industry is not pursuing it because it is the victim of bad philosophy. A programmer with the wrong epistemology will eventually fall prey to it, but armed with a good one, he is *unstoppable*. Thus, I guess that programmers will continue to be significant contributors to the human project, only more so after the invention of AGI.

Where does current AGI research go wrong and how might one build it? How did creativity evolve in humans? Are our computers capable of running AGI? Is it safe, or instead man's last invention? Once built, what might the future hold – would it just make life a little more convenient, or can it take us to the stars?

These are some of the questions this book is intended to answer. It is suited for anyone interested in intelligence and provides insight into the latest AGI research. No previous programming experience or knowledge of philosophy is required. Code samples are given here and there but explained thoroughly for laypeople. Do not worry about understanding the technical parts of this book – there are not many. You should know, however, that you are already a seasoned programmer, even if you have never written a single line of code. You will learn why in this book. We will explore the epistemologies of the philosophers Karl Popper and David Deutsch and apply them to software engineering in hopes that this approach may one day help build an AGI. Indeed, the title of this book is a reference to chapter 8 of David Deutsch's book "The Beginning of Infinity" called "A Window on Infinity." While this book heavily draws from Deutsch's and Popper's work, errors are mine. And while I have tried to reference all source material as thoroughly as possible, I am sure some things have fallen through the cracks.

Software is not just a tool to solve problems, but it shapes the universe and exerts causal power on the physical world around us. Many philosophical problems, including AGI, will not be solved until software engineers adopt the requisite philosophy, and until philosophers realize the importance of software engineering. Likewise, many software-engineering problems cannot be solved without philosophy. Both philosophy and software engineering are two sides of the same coin, and I believe this realization will yield fruit in our attempts to build this amazing thing.

1
A Brief Account of the Origin of Computer Programs

Before I can explain my theory of the mind, and with it, how we may one day build AGI, I must first introduce you to the concepts of computation, knowledge, and universality. The first few chapters are my attempt at a summary of these topics as the greatest minds in the corresponding fields, i.e., Karl Popper and David Deutsch, have explained them, with some of my thoughts (and, surely, errors) added in.

To "*compute*" is often used synonymously with "calculate," but that is misleading. This confusion may be responsible for the popular myth that those hoping to become programmers need to be good at math, which in turn deters many from pursuing the craft. While computers did indeed inherit their name from people who manually compiled logarithm tables, they do something far more profound than that.

Deutsch explains the concept of computation in an interview (2014b): "A theory of computation [...] is the theory of how you can use physical objects to represent abstract objects." One does not need to build a machine to do this – as he explains in the same interview, if you want to count to three, you can use fingers to do so. In this example, every finger represents an integer. (*All* thinking is computation (Deutsch 2012a, 186), but I will get to that later.)

We loosely model our present-day computers after *universal computers*. A universal computer takes this ability to represent abstract objects using physical ones to the universal level. It can simulate *any* computable abstraction in arbitrarily fine detail, and all universal computers share this repertoire (that is the kind of universality referred to in the name). Via those abstractions, a universal computer can simulate any computable physical process (Nielsen 2004). When we speak of a simulation, whatever is being simulated is not being "faked." The simulation results in the same information processing (Deutsch 2018a).

The set of all possible motions in a universal Turing machine, which is equal to the programs that one can run on it, "is in one-to-one correspondence with the set of all possible motions of anything." (Deutsch 2014b) So when a software engineer writes and then runs a computer program, he instructs that computer to move its internals in precisely such a way as to simulate a particular set of abstractions. He instantiates "abstract objects and their relationships in physical objects and their motion." (Deutsch 2014b) Therefore, writing programs is always about simulating abstractions and finding those instructions that will cause the computer to move its internals physically as is required. In a way, software engineers are physicists.

Why do software engineers write programs? Programming, like any creative endeavor, always starts with *problems*. Perhaps one of the first programming problems was how to automatically calculate logarithm and cosine tables for use in navigation and other areas. Large numbers of people – "computers" – compiled these tables. They contained errors, which cost lives. If this process could be automated using a machine, fewer errors might find their way into the tables.

A tentative solution to this problem was the Difference Engine. It was first conceived of by the German engineer Johann Elfrich Müller in the late 1700s, and again by the English polymath Charles Babbage to address the problem of dying seafarers in the early 1800s. Unfortunately, neither of them ever built it. Later on, Babbage envisioned the first *universal* computer, which he called the "Analytical Engine." The English mathematician and associate of Babbage's, Ada Lovelace, even wrote one of the first computer programs.

A problem is not necessarily negative. It is a conflict between two or more ideas (Deutsch 2012a, 31). Babbage recognized conflicts between the actual and desired accuracy of logarithm tables used for seafarers; and also, on a more tragic note, between the desire to save their lives and the reality of their deaths. Today, we write programs to solve all kinds of problems: instant messengers solve the conflict between the desire for real-time communication and the reality of the slow speed of snail mail. Facial recognition solves the conflict between the need for faster identification and the reality of clunkier mechanisms of identification, such as usernames. The *universality of computation* implies that one can solve all soluble problems by writing the requisite software.

While developers write all of their programs to solve problems, merely *solving* a problem is not enough. Not *every* solution will do. It has to be a *good* solution: a good program implementation. There is an objective difference between a good and a bad implementation: a program has a good implementation when it is *adapted* to solving the problem it purports to solve. For it to be adapted means that few changes would make it perform better at solving its problem, and most changes would make it perform worse at that purpose. Every part of the implementation plays a vital role in solving the problem, and changing any part would break its ability to do that. In other words, the program is *hard to vary* "while still accounting for what it purports to account for." (Deutsch 2012a, 31)[1] This means a good implementation resists change. All popular software-engineering principles, such as modularity and reusability, are special cases of the globally applicable principle of being hard to vary.

The origin of every program is the programmer's *creativity*. To solve a problem, he must first conjecture a solution. Afterward, he can try to improve it. He does this many times and conjectures and criticizes many different approaches in an effort to eliminate bad ones. We know of no other way to do solve problems (Popper 2002b, 74). This process is *evolution*: trial and error correction (Popper 1983, 256-284). The philosopher Karl Popper also referred to it as conjectures and refutations (though he did not only apply this to software engineering but knowledge creation in general). A programmer guesses solutions, criticizes them, and picks the best remaining one – if there is one. If there is not, he has to guess more candidate solutions and criticize them once more, until he has found a tentative solution. That is creativity, and it fuels all human problem solving and knowledge creation, not just software engineering. Since a good program implementation is literally the result of evolution, it is no metaphor to consider it well *adapted* to solving a problem.

A *bad* implementation is *easy* to vary while still accounting for what it purports to account for. It may solve a problem, but it does not solve it well. The worse the implementation, the easier it is to make improvements to it, though this always requires additional creativity. Being easy to vary can go two ways. Either a program is easy to vary *in*ternally, meaning parts of the program can easily be omitted or changed to improve it, or it is easy to vary *ex*ternally, meaning it is not adapted to any particular purpose. It does not solve any particular problem, which can sometimes mean it solves several problems poorly.

Why choose good implementations over bad ones? Because there is a truth of the matter about what constitutes a solution to a problem. As such, programming is an effort to create explanatory knowledge, which is a continuation of an age-old cosmological endeavor: the project of understanding reality, which is unique to people. When there is a problem, meaning a contradiction between two ideas, it tells us that at least one of them is false, because, in reality, there is no such contradiction. Hard-to-vary programs are preferred because choosing one of countless variants without any functional advantage is irrational (Deutsch 2012a, 21). Since a good solution to a problem has sufficient explanatory power to explain everything the conflicting theories do, it usually has a conserving as well as unifying character (Popper 1983, 369). But do hard-to-vary programs contain *truth*, or are they merely useful? The answer is the former, and I will examine the connection between programs and explanatory knowledge in more detail in chapter 3.

Good programs have the interesting property of contributing causally to their *replication*. First, they will come out on top of rival programs during the programmer's attempt to solve a problem: it is against his interest to discard good implementations. Moreover, they will spread across computers and gain popularity. A good implementation that solves real-world problems will, for example, find users by spreading across the internet as a website or as an app that one can download, and users who like it will tell others about it and encourage them to download it, too. This way, the program ensures[2] that it keeps running. It is a *replicator*. Social networks are a prime example of this effect. Moreover, since replicators are one of the primary ingredients of evolution, the evolution of programs is already happening across computers. As I will discuss, building an AGI requires replicating this effect *within* a computer.

I should point out that when I say "evolution," I am not only referring to a *biological* concept. For, evolution is fundamentally a theory about abstract replicators. (This fact is reflected in what is now called "neo-Darwinism.") Genes merely happen to be instances of such replicators. Nonetheless, the origin of computer programs does closely resemble the origin of life (more on that in chapter 5). Evolution has so far only happened in a minimal sense on computers, in the form of "evolutionary algorithms." They are limited because, like all other computer programs, they reliably carry out a particular purpose and then terminate. In a sense, they are pessimistic: they *only* solve a problem or set of problems without finding new ones. Current computer programs do not make mistakes, but mistakes are a vital ingredient of creativity because they make error correction possible in the first place. Thus, present-day programs resist change and cannot keep growing in an unbounded fashion. To facilitate unbounded growth, we still require the programmer's creativity. AGI, on the other hand, will be able to make, find, and correct mistakes and keep solving ever newer and better problems in an unbounded fashion just like people do.

Terminology

To compute – To instantiate "abstract objects and their relationships in physical objects and their motion." (Deutsch 2014b)

Problem – A conflict between two or more ideas (Deutsch 2012a, 31)

Computer program – A set of instructions that command a computer to move its internals in a way that instantiates the desired abstractions

Good/bad program implementation – An implementation of a program that is hard/easy to vary internally and/or externally while still accounting for what it purports to account for

Well/poorly adapted to a purpose – Being hard/easy to vary while fulfilling that purpose

Internal variability – When it is easy to change the internals of something without breaking its ability to fulfill its purpose

External variability – When something is not adapted to any particular purpose

Replication – Self-induced copying

Replicator – Something that contributes to its copying

Creativity – The ability to create new explanations

Evolution – The creation of knowledge through alternating variation and selection

AGI – Artificial General Intelligence; a running instance of the creative algorithm

Summary

Programs originate in the creative minds of programmers who write them to solve problems. They instruct a computer to move its internals in such a way it will instantiate abstractions that will solve the problem at hand.

Programs are the result of alternating conjecture and criticism in the programmer's mind: evolution. They replicate, meaning they contribute causally to their copying. A good implementation is hard to vary internally and externally.

Once programmers create knowledge to solve a problem, they can make this knowledge explicit by translating it into computer programs. Now, how do programmers – or anyone, for that matter – know anything? What is knowledge, and where does it come from?

2
How Do We Know?

The question of how we know is an old problem. Our best answers to this question were given by Popper (1983, 2002b) and Deutsch (2012a). This chapter roughly summarizes their findings.

The problem of how we know arose in combination with the problem of change.[3] Philosophers in ancient Greece noticed that some changes escape our senses, and yet we know that they occur. "We do not *see* our children grow up, and change, and grow old, but they do." (Popper 2002b, 194) This realization is surprisingly simple yet profound. It hints at how limited a role the senses play in the creation of *knowledge* and how they can sometimes even hinder it: if we focused only on what we *see*, we would never notice our children growing up. That requires additional conjecture.

To solve problems, we need *explanations*. They answer questions of "why?" and "how?". In doing so, they describe reality: they explain what is really out there, how it works, and why (Deutsch 2012a, 30). So knowledge is not just an accumulation of statements or facts – for example, a history book containing only facts about a war and the corresponding dates by no means explains the war's causes or its lasting effects.

We create explanations when we solve problems. As I said previously, a problem is a conflict between two or more existing explanations; for example, when a theory makes a prediction that is expected to come true but does not. The result is a conflict between the interpretation of an observation – also a theory – and the original explanation. For example, a conjurer may perform a trick that makes it look as though a ball is floating in mid-air. Such a trick would violate one of the more basic explanations in your mind about the world: your best (mostly intuitive) explanation of gravity says that the ball should fall to the ground. The *problem* is that you do not observe it falling. You will want to explain it. To do so, you require a new explanation that does not have the conflict, and yet still explains everything the conflicting theories explain. So the new theory (the solution) has to conserve the previous theory's explanatory power, plus it needs to explain the problematic observation.

One viable – though in this case not very detailed – explanation is that the conjurer is performing a trick and that the laws of physics and gravity still hold. As I stated in the previous chapter, the reason we look for explanations that do not have the conflict is that our theories are accounts of how reality works, and no such conflicts exist in reality. Therefore, any conflict between theories shows that at least one of them is false. Sometimes, all of them are.

All this is as true for magic tricks as it is for the inner workings of stars, the mechanisms of curing a disease, and how to write computer programs. We need explanations for all of them. Where do explanations come from? For much of history, people thought that we derive them from our senses. This idea is known as *empiricism*. However, empiricism cannot be true, because we could not possibly derive knowledge about empiricism itself from the senses, and so empiricism rules itself out. Though a simple and powerful refutation, empiricists ignore it; they prefer to marvel at the mysteries of how we receive knowledge from our senses anyway, instead of looking for a better theory.

According to our best theories, we *guess* knowledge. The origin of all human knowledge is bold conjectures. We create new knowledge by "rearranging, combining, altering and adding to existing ideas with the intention of improving upon them." (Deutsch 2012a, 4) This activity is creativity.

Senses do play a role – but only when we test our theories, and tests are only a small part of the criticism to which we can subject our theories. Because theories are the result of guesswork, we should only ever adopt them tentatively. All people make mistakes – we are fallible – so we should expect even our best knowledge to contain mistakes in addition to truth. There are no authoritative sources of knowledge, nor is there a way to establish a theory's truth or likelihood. The acknowledgment of these facts, which acknowledgment is a key ingredient of Popper's epistemology, is known as *fallibilism*. We should always expect to find more problems with our theories, and even better explanations to supersede those theories. As long as we continue to look for problems, this process can continue forever. This way, we can make unbounded progress, and that is why science and philosophy are both unended quests.

Knowledge is adapted information (Deutsch 2015, 5:06). Just like computer programs, it is information that is adapted to the purpose of solving a problem. Only few changes would make it perform better at that purpose, and changing it would be difficult without making it perform worse.

When Popper worked on finding the demarcation between science and non-science, he suggested that for a theory to be considered scientific, it needs to make testable predictions. But testability alone cannot be sufficient (Deutsch 2012a, 14). Consider the following bad explanation of lightning: If somebody told you that the Greek god Zeus caused thunder by angrily throwing bolts of lightning down onto Earth from up on Mount Olympus, would you consider it a *scientific* theory? It *explains* the cause of lightning in the sense that we can deduce the occurrence of lightning from it. It is testable as well: it predicts that if we climbed up Mount Olympus, we could see and meet Zeus there; presumably, we could even talk to him. But no-one in his right mind would send an expedition.

Popper understood this. For example, he said that origin myths, which often invoke gods and other supernatural entities, whimsical as they may be, are fine starting points to explain the origin of the world as long as we are willing to ask some awkward questions (n.d.).

What is the purpose of awkward questions? It is criticism, which almost always precedes tests. We can criticize the Zeus theory of lightning because it is too easy to change some of its details without destroying the entire theory. For example, why Zeus, and not a wizard? Why is he angry? Why is he on Mount Olympus and not another mountain? Greek myths may have placed him there, but the theory still works if he throws lightning off of other mountains.

Just like bad programs, this explanation is easy to vary. We can easily change its components without hurting its explanatory power. This explanation, by definition, is not adapted to the purpose of explaining lightning. Therefore, it is a bad explanation and not scientific. Consider the following variant, which is just as bad: Whenever Zeus is especially *happy* with humans, he celebrates them by causing a lightning show similar to today's fireworks. The reason both explanations are so easily variable is that their details have no bearing on whether the phenomena we are trying to explain will occur (Deutsch 2012a, 21): it does not matter if Zeus is happy or angry, he may cause lightning either way. There is no reason to prefer either theory because they both make the same predictions while being easily variable, and they allow us to change their internals to account for problematic observations should we need to do so.

Ultimately, all bad explanations are variants of a single, very bad explanation, that is both easy to vary internally as well as externally: "the gods did it." (Deutsch 2012a, 21) It is easy to vary *internally* because you can replace gods with any other supernatural being, such as angels or wizards, and it still explains whatever it purports to explain. Likewise, it is easy to vary *externally* because it accounts for any problem whatsoever: it explains anything at all. When we refute an explanation, it becomes easy to vary: there is a problem it cannot account for, making it easy to vary while still not accounting for that problem. Through the ages, there have always been charlatans. Today, they include advocates of homeopathic "remedies," chiropractors, acupuncturists, and the like. Thankfully, we can quickly identify them by the bad explanations they propose.

Our best explanation of lightning is hard to vary: roughly, it explains lightning as an electric current that is the result of electrical charge building up in a cloud due to friction between ice crystals. High altitudes are sufficiently cold for ice crystals to form. You will fail to change the internals of this explanation without it falling apart. Remove the ice crystals, and you do not have friction. Without friction, you do not get the necessary electrical charge. And without connected theories of weather and our atmosphere – all hard to vary in their own right – you do not get clouds at all. This theory resists change. That is also why when one person tells it to the next, it is transmitted faithfully, even though neither party necessarily cares about its faithful replication (Deutsch 2012a, 257).

This good explanation of lightning has what Deutsch calls *reach* (2012a, 28). It transcends our atmosphere and predicts that we should find lightning on other planets with similar atmospheres. We do: on Jupiter as well as Venus (although interestingly, Venusian crystals are made not of water but sulfuric acid). Any explanation involving Zeus has limited reach. If he is on Mount Olympus, he can presumably throw lightning only far enough to cause it in Greece, unless he has superhuman throwing abilities to cause lightning on the other side of the planet, which is another bad explanation.

In 1950, the mathematician and historian Jacob Bronowski was asked to examine the teeth of the two-million-year-old fossil skull of a newborn Australopithecus, a member of a group of extinct African apes, in order to mathematically distinguish their size and shape from the teeth of other apes. He reflects on that time: "I had never held a fossil skull in my hands, and I was by no means an expert on teeth. But it worked pretty well, and transmitted to me a sense of excitement [...]: I, at over 40, having spent a lifetime on doing abstract mathematics about the shapes of things, suddenly saw my knowledge reach back two million years, and shine a searchlight into the history of man. That was phenomenal." (Bronowski 1973) This is the kind of reach a good explanation can have.

Bad explanations can also have reach, but this is what I have described as external variability. They are not adapted to solving any particular problems, and so they are poorly adapted to solving a variety of problems. What is remarkable about *good* explanations with reach is that they are *adapted* to solving a multitude of problems, which makes them more vulnerable to criticism and therefore promotes the growth of knowledge.

A theory that immunizes itself against criticism by being easy to vary may survive over time. Therefore, one should always be suspicious of traditions and theories that have not changed over long periods of time despite being easy to vary.

As I argued in the previous chapter, being hard to vary is a desirable feature of explanations because, when an explanation is easy to vary, choosing one of countless variants without any functional advantage is irrational (Deutsch 2012a, 21). The search for good explanations also implies a *criterion of reality*: "we should conclude that a particular thing is real if and only if it figures in our best explanation of something." (Deutsch 2012a, 23)

The origin of all knowledge is repeated conjecture combined with repeated criticism: creativity. The conjectural nature of knowledge was known to Greek philosophers from Presocratic times. As Xenophanes wrote (Popper 2002b, 205):

> The gods did not reveal, from the beginning,
> All things to us; but in the course of time,
> Through seeking we may learn, and know things
> better. […]
>
> But as for certain truth, no man has known it,
> Nor will he know it; neither of the gods
> Nor yet of all the things of which I speak.
> And even if perchance he were to utter
> The perfect truth, he would himself not know it;
> For all is but a woven web of guesses.

Even the limited role of observation, and hence the invalidity of empiricism, was known then. Democritus wrote (Popper 2002b, 206):

> But in fact, nothing do we know from having seen it; for the truth is hidden in the deep.

Xenophanes and Democritus, among other philosophers at the time, were interested in *epistemology*. Though they could not have stated it this way, they were working on AGI long before the invention of computers.

Why was this approach to knowledge abandoned? The smoking gun points to the inventor of *induction*: Aristotle. While thinkers before him distinguished between certain knowledge ("epistēmē," which regrettably still lends its name to the word "epistemology") as only accessible to gods, and opinion ("doxa"), which mortals can access and improve through guesswork, Aristotle broke with this tradition. He wanted to achieve certain, demonstrable knowledge; he wanted to see the essence of things once and for all and considered the goal of science to be an ultimate collection of essences. He killed the critical tradition that had come before him, and what survived is still known and falsely desired in much of the scientific community of today as "justified true belief," which is related to both empiricism and *inductivism*.

While empiricism is the misconception that we "derive" knowledge from the senses or experience – "you need only clear your mind and read the book of nature" – inductivism takes this a step further. It is the idea that knowledge is "obtained" somehow by extrapolating *repeated* experiences, and that repeated supporting observations make a theory more likely to be true. Inductivists are after "justified true belief," thereby espousing *justificationism*: they only consider knowledge "real" or "genuine" if it originated in the senses. But an attempt at justification always leads to an infinite regress: if knowledge is justified by experience, or some other alleged source, what justifies that source?

Nobody in their right mind, so the argument goes, would doubt that we derive knowledge from the senses (empiricism) and that we improve it and become surer of it by making repeated observations (inductivism). I suspect that both doctrines spread because they make each other more plausible and because induction is the default epistemological position most people hold.

At first glance, induction does seem to solve the problem of the origin of knowledge, in addition to serving as a criterion of reality: we do not, so the theory goes, consider something real unless we have made sure of it through repeated observation. It even offers a *criterion of demarcation*: we do not, so the argument goes, consider something scientific or reasonable or rational or whatever variation thereof unless it can be "proven" or "verified" by observation. Therefore – ironically – it is especially scientifically minded people who want to protect inductivism as their epistemology. Without it, many think, we must conclude that gods and invisible pink elephants and teapots orbiting the sun (as Bertrand Russel imagined) are all real!

But we fallibilists are not trying to remove induction without replacement. The attempt to replace it will incur some ridicule, but no matter. We have something at our disposal that can solve all problems induction purports to solve without running into the infinite regress of justificationism or having to explain how repeated observations produce knowledge. We understand that explanations are the result of alternating guesses and criticism. The quest for good explanations provides an adequate replacement for both criteria. As a criterion of reality, it says that we consider something real if it figures in our best explanations. As a criterion of demarcation, it says that we do not consider something scientific unless it is a good explanation that is falsifiable by observation.

We can, therefore, safely dispose of inductivism without having to grant the existence of gods, invisible pink elephants, or teapots in space. And it is necessary to do so, for, in reality, induction does not exist, and the goal of science and philosophy is unbounded problem solving, not having an ultimate collection of anything. Induction cannot exist for the same reason software engineers do not observe or "read" their programs off of nature, and do not improve their programs by repeating those observations: most of our knowledge does not involve experience. We know how hot stars are, even though nobody has ever touched them. We know how dark and cold intergalactic space is, even though nobody has traveled there yet (though we may soon; see chapter 10). We know of *abstractions*, such as numbers and moral ideas, even though we cannot possibly experience them. They are not physical entities, and so no light could bounce off of them to enter our eyes. We do not derive knowledge from observation or repeated observations; that is not how we learn. Nor could repetition on its own be sufficient to create new knowledge because it includes repetition of errors.

Instead, we learn by *reasoning*: by critically assessing conflicting theories and conjecturing solutions to problems. It is necessary to dispose of induction not only because it is false as an explanation of the origin of knowledge, but also because it is a harmful theory that has stifled progress and introduced more suffering than most other bad explanations (see chapter 9). It is difficult to overstate how bad an idea induction is. Popper tried to eradicate it; alas, he failed.

As I have said, many scientifically minded people are ironically protective of inductivism. In addition, many scientists have an aversion to philosophy. Feynman famously said that "Philosophy of science is about as useful to scientists as ornithology is to birds." But that is wrong, because, while ornithology merely provides explanations and descriptions *of* birds, philosophy of science not only explains how science works but explains, *to scientists*, how they *ought* to do it if they want to have a chance at success.

Now, most philosophy of science is indeed useless because it is entirely false and pointless chin-stroking. Many scientists, therefore, correctly feel that philosophy does not get anywhere. If instead they focus on scientific matters in front of them, they do make progress, sometimes fairly quickly. That may be so, but whatever methodology they follow when they do science *is* their philosophy of science. You cannot do science without it. How to do science is not a scientific matter because we cannot refute it by experiment.

Of course, whenever scientists – or others – *do* make progress, it must be due to error correction, which can only have happened if they temporarily abandoned their flawed philosophy of science, perhaps without realizing it. Popperian philosophy – critical rationalism – is a philosophy of error correction, which is why it is the one philosophy that does make progress.

Imagine a scientist who is an inductivist. This conviction informs how he does science. Given that inductivism is wrong, would he not want to know? With the wrong philosophy, he will make mistakes *in science* all the time, and he will have no way of detecting or correcting those mistakes without a good philosophy. He will keep making them, unaware!

In any case, the empiricism of the early Enlightenment was a useful defense against the authoritarianism and superstition of the middle ages, when people attributed knowledge to divine sources, and believed to know everything significant already (Deutsch 2012a, 5-6). New atheists such as Sam Harris still use empiricist reasoning to refute the existence of god by asserting one should not believe claims for which there is no evidence. But unfortunately, empiricism only replaced divine authority with the authority of the senses (Deutsch 2012a, 8-9). It continued the mistake of judging knowledge by its source. The apparent source or origin of knowledge is irrelevant. It does not matter *who* tells you something, or if an idea comes to you in a dream or while you are awake (Deutsch 2012a, 223-257). Knowledge either solves a problem, or it does not. Its source is always the same: your creativity. And the real reason one should reject god or any supernatural being as an explanation for anything is not that we have no evidence of them, but that they are bad explanations.

Conversely, one should accept explanations without evidence if they are good. Imagine reading a newspaper article that claims that there is no evidence that flossing has any advantages for your dental health and concludes that you do not need to floss. That is irrational because we have a good explanation of why flossing works: it removes food particles from between your teeth, which are places that may be difficult to reach with a toothbrush. If these food particles are not removed, bacteria from such particles can remain on your teeth longer and cause problems such as tooth decay and plaque buildup. No evidence is needed to prove this explanation true. No evidence *could* prove this explanation true. Thinking that it could is an empiricist mistake. It is a good explanation, so floss. Only stop flossing if the explanation is refuted.

How could it be refuted? One way is to find a problem with the explanation, such as an inconsistency or contradiction. Another is to find a conflicting explanation that is better. Or, an interpretation of an experiment could find that flossing teeth leads to worse dental health.

In the 20th century, people started to take empiricism seriously, which almost led to the death of some sciences (Deutsch 2015, 14:55). Aware of the shortcomings of empiricism, many scientists and philosophers concluded that science cannot tell us anything about reality, but can only make predictions. This idea is known as *instrumentalism*, which is one way of denying reality and the systematic refusal to ask "why?", that is, to seek explanations, and it is false because there cannot be such a thing as an "explanationless prediction." (Deutsch 2012a, 15)

In a lecture about the scientific method, Popper (n.d.) explains that Aristotle was the first philosopher who limited himself to writing only about those problems he could solve. So did the Greek mathematician and father of geometry Euclid, who succeeded in solving several cosmological problems Plato had run into and then compiled these solutions into his "Elements." Alas, claimed Popper, he made two mistakes, the first of which we have somewhat remedied in modern computer programming. It is that he did not state what he was doing: which problems he was solving and why. Perhaps, Popper says, Euclid did not see the need because everyone at the time knew he was a Platonic cosmologist. I say that we have *somewhat* remedied this in computer programming because we *always* write programs to solve problems (as opposed to some of the material created in other fields), and these problems are often, though not always, manifest in the resulting implementations. Of course, a programmer may nonetheless regularly find himself wondering what on earth the particular piece of code his esteemed colleague wrote *does*. That is always a question of what problem the code is adapted to solving because, as Popper often emphasized, understanding a solution involves understanding the problem situation from which it originated.

Additionally, Popper claims Euclid made a second mistake – Aristotle's mistake – when he wrote in his "Elements" only of those problems he had solved, and left out the ones he had not. "This, in a certain sense, was a crime," so Popper. It effectively "put an end to geometrical research" and started an intellectual tradition of solving a problem without stating new or remaining ones, even though they inevitably occur. But it made for a good "ultimate collection" of mathematical facts, so nobody bothered grappling with the conjectural nature and outstanding problems of Euclid's "Elements."

I guess this tradition of writing only about the problems one has solved still has far-reaching implications today and that our present way of writing computer programs is a direct consequence of it. It is why I said in the previous chapter that our programs are pessimistic: they solve a problem and then *terminate*. They do not make mistakes or find new problems and can, therefore, never be open-ended problem solvers: AGIs. But why should our present way of writing programs be the only way? Had the designers of the first programming languages been aware of this traditional mistake, we might already have AGI: it could almost have fallen out of the way we would have naturally programmed.

Karl Popper re-discovered and re-introduced the critical attitude while working on a solution to a problem with induction known as *the* problem of induction: if, say, all so far observed swans are white, one cannot validly conclude that *all* swans everywhere are white. Conventional wisdom said that the future will resemble the past. Often, that is simply not the case. *Things change.* Moreover, nobody has ever been able to formulate a principle of induction that enables extrapolations of the kind upon which induction relies.

Popper solved this problem by realizing that people reason not through repeated observations, but by solving problems through alternating conjectures and criticism. He made the most significant contributions to the field of AGI – though, like Xenophanes, he presumably did not consider himself an AGI researcher. But that is what he was. He restored the concept of conjectural knowledge and open-ended problem solving and rediscovered some crucial laws of epistemology along the way. The creation of knowledge does not start with observations, but with problems: conflicts between explanations. We solve problems by *explaining* them. We never observe an explanation: it is something unseen that accounts for the seen (Deutsch 2009). It is this concept that is vital to progress in science and philosophy as a whole, and to the building of AGI in particular: an AGI is a program that can create knowledge just as any other person can; it can solve problems by *explaining*.

The creation of human knowledge is a continuation of a biological phenomenon. Science, philosophy, software engineering, and indeed all of human creativity is a continuation of *biological evolution* (Popper 1983, 256-284). Since knowledge does not require a knowing subject (Popper 1983, 106-152), all kinds of media can store it. Knowledge in the form of genes is information that is adapted to solving the problem of how to spread the gene through the population at the expense of its rivals. Both human knowledge and biological adaptations are hard to vary and contribute causally to their replication: (ideas that replicate across people are called "memes" (Dawkins 2016, chapter 11); more on these in chapter 9).

Although it happens much more rapidly in human minds than in nature, the process of trial and error correction explains not only the origin of human knowledge but also the origin of species (Popper 1983, 256-284). Genes are replicators, and they encounter problems, too: changes in the environment can make it harder to spread through the population and suppress rival genes. A genetic mutation can be an attempt at solving that problem. By "attempt," I do not mean to attribute any agency: a mutation occurs by chance, through blind variation, and without regard for the problem situation. It always only *happens* to solve a problem and appears goal-directed only in hindsight.

From the perspective of the gene, problems are not just about replication per se: they are about how the gene can create, and turn into, the thing that will make it spread through the population. Whenever an organism encounters a problem, its genes can spread if it solves the problem by invoking knowledge stored in them. Absent such a solution, it will die. In that case, its genes can only spread if they are also present somewhere else in the gene pool. But they must have mutated sufficiently to incorporate a solution to the problem. Those variations that happen to solve the problem may spread, while those that do not go extinct. That is how evolution occurs.

Most genetic mutations are fatal: they lead to the death of their holder. Conjecture has its counterpart in chance variation, and error correction has its counterpart in the death or elimination of a replicator. Likewise, many epistemological misconceptions have counterparts in misconceptions about biological evolution (Popper 1983, 256-284). We should expect this, for , as I have said, evolution is fundamentally not about anything biological, but about abstract replicators. Therefore, misconceptions about biological evolution may well apply to other types of evolution, such as the evolution happening in our minds.

Before Charles Darwin developed the theory of evolution, the French naturalist Jean-Baptiste Lamarck offered a false solution to the problem of the origin of species, known today as *Lamarckism*, in his 1809 book "Philosophie zoologique." The theory states that an organism could inherit physical characteristics that a parent organism had acquired during its lifetime, and that such characteristics would develop and disappear through use and disuse. For example, a giraffe may stretch its neck to get to leaves hanging higher up on a tree, allegedly leading to a slightly longer neck, which its offspring is said to inherit. This process would repeat over generations, leading to the long-necked giraffes we know today.

Though mistaken, Lamarck should receive credit for proposing a theory of evolution that is driven by natural rather than miraculous processes. Much like empiricism, Lamarckism did its job shaking the prevailing world view just enough to pave the way for a better system. Lamarck's theory was appalling even to other naturalists who believed that nature was evidence of the Christian god's benevolent design. He was rejected for his heresy and died poor and forgotten in 1829.

Changes to an organism can only occur during its lifetime if it already contains genetic adaptations causing such changes. So Lamarckism does not explain how that knowledge was created in the first place, which is what it set out to do. Lamarckism and induction share the mistake of claiming that knowledge is somehow already present in experience, or that one can derive it from repeated experience or use and disuse (Popper 1983, 256-284). In reality, chance mutations create new knowledge *first*, without regard for the problem situation, and natural selection discards those mutations that do not solve it. So both the human and the genetic kind of knowledge need to be conjectured first before being tested. That is what both Darwin's theory of evolution and Popper's theory of knowledge creation say.

Biological evolution offers a further refutation of empiricism: before an organism can consume any sense data, its predecessors must have evolved sense organs which it inherits. Those organs are conjectured solutions to problems. Throughout their evolutionary history, genes encoding their organisms must actively "reach out" (again, no agency implied) into their environment to find out about its structure. They do this using conjectured tools of exploration such as sense organs, combined with the conjectured theories they embody about how to interpret the incoming sense data. So the theory always comes first; before any observation can be made. The environment does not instruct organisms.

Contrary to human knowledge, the knowledge stored in genes is almost always limited – so limited, one may consider it non-explanatory. Genes only encode knowledge of how to, say, fly in order to spread through the population, not in order to help the organism fly. Therefore, even an organism that can fly is not *adapted* to the purpose of flying. If it was, its genes have had enough time in evolutionary history to evolve its abilities well beyond those of airplanes and rocket ships. Instead, any organism's purpose is always limited to helping spread its genes through the population.

In addition, genetic knowledge has limited reach because it is always the same solution to a single problem: it explains how to spread the gene through the population at the expense of its rivals. It only contains *approximations* to explanations (such as how to fly).

Lastly, as opposed to human knowledge, genetic adaptations need to work every step along the way: almost all mutations in genetic knowledge lead to the organism's death, or at least serious disadvantages compared to rival genes. That's what it means for a gene to be well adapted. If an organism with a mutation dies before it can reproduce, the variation does not get a chance to mutate even further. Human knowledge, on the other hand, does not have to withstand this pressure. We can alter an idea and, acknowledging its deficiencies, go through a "valley of bad explanations," that is, keep altering it in the hopes that it will eventually bring about a good solution to a problem, even if along the way changes temporarily make those ideas worse. (While we can intentionally alter an idea, the process of finding a better idea always seems goal-oriented only in hindsight, for if you knew the "target" idea at the beginning, there would be no need to evolve it through thought.)

In other words, human knowledge has more leeway in finding explanations, which can frequently lead to explanations of great, sometimes even unlimited reach. Reach is much rarer in genetic knowledge. It is this leeway in addition to an explanation's reach that allows us to almost effortlessly imagine creatures with fantastic mutations such as Pegasus: a winged horse capable of flight. Such mutations are much harder to come by through biological adaptations.

A related difference between human and genetic knowledge is that while disadvantageous mutations usually lead to an organism's death, a bad idea does not lead to death per se. An organism has no choice but to dogmatically hold on to a harmful mutation it inherited and die with it. People, on the other hand, can entertain an idea and, if they find it to be inadequate as a solution to a problem, they can discard it without dying. Popper liked to explain this by considering the difference between Einstein and an amoeba (1983, 24-25). When an amoeba procreates, and variation alters the knowledge in its genes for the worse, its offspring will die. But when Einstein has a new idea, he can explore it, and, if he judges it unviable, he can simply discard it without dying himself. The genes in the amoeba avoid change because it may cause death, whereas Einstein embraces change as it can lead to better ideas. As Popper said: we can let our ideas "die in our stead." (1983, 244)

Genes effect changes in the physical world. One such change is the use of material for their replication. Another is to build a protective suit of armor – bodies – around themselves, and determining the specific parameters of that body, such as eye color, the functionality of the heart, and so forth. Some genes have enough reach to cause changes outside their own body: for example, through beaver dams and spider webs (Dawkins 2016, chapter 13). The set of changes a replicator causes in the physical world is known as its *phenotype* (Dawkins 2016). DNA molecules are part of the genetic replicators' phenotype. (I will revisit phenotypes in chapter 5.)

Since genes and people create theories in response to problems, and since problems are born out of mistakes in existing theories, it is interesting to think about what the very first problem and the very first theory may have been. To solve the circularity, they must have arisen simultaneously. I guess that the first problem had to do with the question of how to replicate faithfully and that the first theory was a theory about faithful replication, which was created by the precursors to life as we know it today.

We do not yet know the origin of genetic replicators, and, with them, the origin of life, but there is a good candidate explanation involving *catalysts*. A catalyst is something that causes an event or change without undergoing any net change itself. Billions of years ago, when the surface of the Earth had cooled sufficiently for water to condense, oceans emerged. These oceans contained molecules. Some molecules formed spontaneously, while others were created with the help of *chemical* catalysts. At some point, one such catalyst facilitated the production of molecules of which it consisted. It was not yet targeted enough to be considered a replicator, but now there was a *chance* for one to be created because the catalyst created variants of itself. Over time, catalysts emerged that created ever more of their components until at least one of them was targeted enough to be considered a replicator. Such a catalyst can then be considered *autocatalytic*. Because evolution favors replicators with high copying fidelity, ever more targeted replicators evolved, until the emergence of DNA. (The idea that the first autocatalytic molecules were made of RNA is known as the "RNA-world hypothesis.")

That is why I said the first problem was the problem of how to replicate faithfully. It was both introduced and solved by the first replicators. Since, again, evolution is a theory about replicators that explains the origin of all knowledge, not just biological knowledge, we should expect the creation of AGI to require something similar to the origin of life as I have described it. (I will elaborate on this in chapter 5.)

Let us consider a real-life problem: say, you own a car and find your parking spot to be empty. The problem is a conflict between explanations: one explanation says that you parked your car in your parking spot last night and predicts it should be there today. An interpretation of sense data constitutes the conflicting explanation that says that the car is not in your parking spot. What you need to do is conjecture an explanation that does not have the problem. It needs to explain how it is that you cannot find your car in your parking spot even though you parked it there last night.

Maybe it was in the way of something and had to be towed? This conjectured theory makes a prediction: you should be able to call the local towing company to see if they have your car. If you call them and they do not, the theory is refuted, and you need to conjecture another theory. Maybe the car was stolen. This scenario is cause to be upset, and you call your best friend to tell him about what happened. What he needs to do as he is listening to you is to replicate your theory based on the limited information you give him. By doing so, he actively creates knowledge, which requires creativity. That is why so-called "recipients" of knowledge are never passive; they have to create actively. He now has to recreate the explanation you conjectured for the visual of the missing car by different means: by listening to your voice. All you and he ever get are clues as to what may have happened; something unseen happened that you need to explain with the seen. Your friend is in that same boat when you tell him about it: he needs to explain what happened with what he hears. He may ask you questions to criticize his candidate theory and to make sure he understands you correctly. With some luck – although there is no way of knowing this for sure – both you and he will have roughly the same theory in your minds at the end of the conversation. If your theory and his do not match, that is what constitutes a misunderstanding.

It may appear as though your friend is in a slightly different situation than you were when you first saw that your car was missing. When you created the knowledge of what may have happened to your car, you had to create it anew from clues in the environment. As your friend is listening to you explain it to him, it appears as though he just needs to hear your words to recreate that same knowledge. But in reality, he, too, can only go off of clues: clues from your words and the knowledge he already has of you and your car. The logic for coming up with an explanation in the first place is no different from the one your friend uses to understand, or copy, that explanation. Creating a new explanation and understanding an existing one both require the same logic.

Popper first guessed this equivalence when he conjectured that the mechanism by which children learn is the same as the one scientists use to create new explanations (Deutsch, personal communication with author, 2019). To understand a story, for example, or even to copy someone else's behavior, we need to explain it – and each of those cases is a "special case of the general human objective of explaining the world." (Deutsch 2012a, 410) That is why building an AGI is a continuation of a cosmological effort. A person trying to understand a story faces the same challenge a scientist does. "Both must discover a hidden explanation. For the former, it is an idea in the minds of other people; for the latter, a regularity or law of nature. Neither person has direct access to this explanation. But both have access to evidence with which explanations can be tested: the observed behavior of people [...], and physical phenomena conforming to the law." (Deutsch 2012a, 411) An engineer is in the same boat when he builds a bridge, or a software program, only from a list of specifications. All of these things are about creating or recreating the unseen from the seen. Whether you are copying or innovating: both follow the same logic.

By reading this book, you are being creative as well: you are guessing what I mean by what I am saying. To understand me, you have to try to replicate my knowledge. You are using creativity right now. You are moving the internals of your brain – also a computer – to instantiate abstractions representing the contents of this book.

To understand anything, we have to replicate that thing in our minds. Understanding something is a *replication effort*, which is another similarity between genetic and human knowledge: both are about replication. That is why I guess that gene replication is fundamentally also about creating the unseen from the seen, namely creating the thing the gene would need to be to spread better through the population, given its current makeup and environment.

The theory of *neo-Darwinian* evolution is, again, a theory about abstract replicators. I say "abstract" because copying any material would involve cloning, so replicators cannot be material, only their instances can. These replicators can be instantiated genetically or in the form of ideas in the minds of people. To understand anything means to be able to explain it, if only to oneself, and that means replicating it mentally. An AGI – a running instance of the creative algorithm – is a universal explainer, which means it is a universal replication system, by which I mean it can create an abstract copy of anything. So where genes only try to create the thing they need to become to spread through the population, people, in their endeavor to explain, mentally try to create the whole world.

Terminology

Epistemology – The study of knowledge: what it is, how it grows, and the laws governing it

Explanation – "Statement about what is there, what it does, and how and why" (Deutsch 2012a, 30)

Empiricism – The misconception that knowledge comes to us through the senses

Induction – The impossible process of deriving knowledge from repeated experiences

Inductivism – The misconception that we create explanations through induction and that theories corroborated by observation are more likely to be true

Problem of induction – One cannot validly extrapolate from a finite set of observations

Justificationism – The misconception that knowledge needs to be "justified by some source or criterion" (Deutsch 2012a, 31)

Instrumentalism – The misconception that explanations cannot describe reality, and that, therefore, one should not even bother explaining anything and only make predictions

Fallibilism – The doctrine that all people inevitably make mistakes and that there cannot be any authoritative or reliable sources of knowledge, "nor any reliable means of justifying knowledge as true or probable" (Deutsch 2012a, 31)

Reach – An explanation has greater/lesser reach if it can solve more/fewer problems in a given domain (see also Deutsch 2012a, 28)

Phenotype – The set of a gene's physical effects on the world (Dawkins 2016)

Lamarckism – The mistaken evolutionary theory that organisms evolve through improved characteristics acquired during an organism's lifetime and inherited by its offspring

(Neo-Darwinian) Evolution – The creation of knowledge through the gradual, non-directed introduction of chance mutations in a pool of replicators, alternating with selection

Biological evolution – Alternating variation and selection of genes replicating in the biosphere

Evolution of explanatory knowledge – Alternating conjecture and criticism of ideas replicating in a single mind

Catalyst – Something that causes an event or change without undergoing any net change itself

Summary

Induction is a myth, as are empiricism and Lamarckism. All knowledge creation happens through conjectures alternating with criticism. Knowledge is information that replicates and adapts. In biological evolution, this happens through the replication of genes: during the replication process, genes occasionally mutate, which can sometimes lead to a gene that spreads through the gene pool at the expense of its rivals. In human minds, evolution occurs through the variation and selection of ideas. In both cases, the result is the creation of knowledge. In human minds, it is explanatory.

Mistakes are inevitable – people are fallible – and there can be no authoritative sources of knowledge. Even our best theories are bound to contain errors, and we can always replace them with deeper, farther-reaching theories. The only thing that offers us a chance of success in any discipline is the quest for good explanations.

To understand something is to create a mental replica of it. We use the same algorithm for creating such replicas as we do to innovate. That is a token of the reach of creativity.

If an AGI is a computer program that creates explanations, we need a way to encode explanations in computer programs. How can we do that? And do the laws of epistemology apply to computer programs, too?

3
Knowledge in Computers – A Functional Approach

The reason we need to study epistemology in order to build AGI is twofold. First, a good program has the same quality as a good explanation: they are both hard to vary while accounting for what they purport to account for. However, this is true of just about anything; a piece of art cannot be good if it is easy to imagine changes to it that would make it better. The same is true of a movie, a book, or a building. The deeper connection and second reason is that if we want to build a program that can create explanations, we need to know how to encode explanations in computer programs. I guess that explanations simply have their counterparts in computer programs and that, therefore, all epistemological laws apply to computer programs. Building on Temple's work on structural epistemology (2003b, 2010) – a field which, according to Temple, originated with Kolya Wolf – I will explore these connections in this chapter.

When I was a boy in elementary school, I remember sitting in math class and listening to our teacher as he introduced the topic of addition. It was probably first or second grade, and I remember being pretty fond of math at the time. Our teacher introduced the topic of addition as follows. Before explaining any of it, he posed an open question to the room: "How do you think addition works?" I raised my hand and said something along the lines of "Well, I know that when people do addition, they say things like '3 plus 5,' so… I guess in order to calculate this, you need to increment three five times. So in this case, it would be 4, 5, 6, 7, 8." In other words, I remembered some environmental clues I had previously interpreted, from which I could guess an explanation of how addition worked. That is knowledge creation in action: my teacher had an explanation of addition in his mind that we as students needed to replicate in order to understand it. We could not download his explanation directly from his mind to our minds: we had to guess instead.

The statement "you need to increment three five times" implies much knowledge. Increment how? Further, it is not five and three for every calculation. So I implied the use of variables here, too, as well as the concept of looping and a condition to terminate the loop. I remember using fingers on one hand to count the number of iterations I had done so far: I would show one finger and go "six," and then I would show a second finger and go "seven," and then I would show a third finger and go "eight." As I was running the addition algorithm in my mind, I used physical objects – my fingers – to represent abstract ones – in this case, numbers. Once the algorithm terminated, I stopped and reported the answer.

Implementing addition as a computer program is a fun and easy way to make this example more explicit. We will use a programming language called Clojure, created by Rich Hickey. It does not matter all that much which programming language we use – most of them are general purpose, meaning you could do what we are about to do with any of them, but Clojure is a remarkably simple language. It will allow us to focus only on the things that matter. As a bonus, much of the research in genetic programming, which is a way of implementing evolutionary algorithms, has been done using this kind of language – Clojure belongs to a family of languages called "Lisps" – so it may help us later. Do not be discouraged if you do not understand everything line by line. All that matters is that you get a general idea.

Consider this implementation of addition:

```
(defn plus [a b]
  (loop [counter 0
         result a]
    (if (< counter b)
      (recur (inc counter) (inc result))
      result)))
```

This program is a *function*. Every program can be written as a function. Do not worry about what the parentheses and brackets mean. Let us go over it line by line:

```
(defn plus [a b]
```

In this line, we are defining a new *function* and calling it "plus." A function is a procedure that we can invoke whenever we want: we only need to specify the procedure once and can then invoke it anytime we need it, rather than writing this logic from scratch every time we want to perform addition. To make this function work for any two numbers[4], we give it two *parameters*, `a` and `b`, which will be replaced with actual numbers when we invoke the function later.

```
  (loop [counter 0
         result a]
```

Here we are starting a loop and setting two variables, `counter` and `result`, to `0` and the value of `a`, respectively. Setting these variables here allows us to update them as we loop.

```
    (if (< counter b)
```

Read: "If the counter is less than `b`."

```
      (recur (inc counter) (inc result))
```

This line only runs if `counter` is indeed less than `b`. In this case, we want to loop again (that is what `recur` does), only this time with `counter` and `result` both incremented by `1`.

```
      result)))
```

This line only runs if `counter` is not less than `b`. In this case, we want to stop looping and simply return the result. The entire function then returns the result. This is known as the function's *return value*.

Consider the following example invocation of `3 + 2`:

```
(plus 3 2)
```

What does this invocation do? First, it sets parameter `a` to `3` and `b` to `2`. Then it sets the counter to `0` and the result to `3`. Next, it checks if `0` is less than `2`, which is true, so it loops again, this time with the counter set to `1` and the result set to `4`. Now it checks if `1` is less than `2`, which is also true, and so it loops once more with the counter set to `2` and the result set to `5`. It attempts the loop one last time and finds that `2` is not less than `2` and returns the result, which is still `5`. The result is correct: `3 + 2` is indeed `5`.

This implementation is fairly well adapted to the purpose of solving addition; most changes will break it. We can change the counter to start at `1` if we also set the condition to <= instead of <. We can also change the order of the parameters to `[b a]` instead of `[a b]` because addition is commutative.

Note how epistemological laws beyond being hard to vary apply to this function implementation. We created it through trial and error correction. Now that it exists, it does not matter who wrote it, and we should not judge its functionality by its origin. Whether we like it or not, it has some reach, contains some mistakes, and carries out a purpose. We do not even need to *believe* in its usefulness or truth-likeness at all. The algorithm is universal for positive integers (it does not work for negative numbers or decimals). It is commutative, as we expect from an addition function.

The function depends on several underlying functions, or "concepts": looping, conditionals, comparison, and increments. That means it would be challenging to build this function without any of these underlying concepts. As a child – if my explanation of addition was anything like this implementation – I would either have needed to know these concepts in advance or come up with them at the time, which would have made the problem harder.

The function is autonomous (see also Deutsch 2012a, 28). What it does is not up to its creator. We may think it does one thing when it really does another. To judge this, we create knowledge about knowledge, which is why criticism and testing are so important. If you criticize this implementation sufficiently, you will find that it does not work for negative integers. To be precise, it does not work whenever the *second* parameter is negative, in which case it always returns `a` because the conditional "if the counter is less than `b`" is never true. That constitutes a problem, which further conjecture and criticism can solve.

Another law of epistemology says that good explanations are discrete (Deutsch 2012a, 342). To get from one good explanation to another, you first need to traverse a "valley of bad explanations." First guesses are almost always wrong and need to be criticized repeatedly to improve because there are many more wrong guesses than right ones. Imagine listing all possible explanations for a given problem in descending order from hardest to vary to easiest to vary (Deutsch, personal communication with author, 2019). Such a list is impossible to come by because it would be infinitely long. Moreover, it would be a source of truth, which cannot exist. Discreteness also implies that picking something halfway between two good explanation will not make a new good explanation (Deutsch 2012a, 342). This reality is a special case of the epistemological law that generating a good explanation always requires creativity and cannot be found mechanistically. It is true for function implementations as well; devising another hard-to-vary explanation of addition different from the one above requires more creativity:

```
(defn plus' [a b]
  (if (zero? b)
    a
    (recur (inc a) (dec b))))
```

What does this implementation do?

```
(defn plus' [a b]
```

This line is the same as in the previous implementation, except we call it `plus'` (read "plus prime") to denote the difference between the two implementations.

```
(if (zero? b)
```

Read: "Is the value of `b` equal to zero?"

```
    a
```

Whenever `b` *is* equal to zero, the function returns the value of `a`.

```
    (recur (inc a) (dec b))))
```

Whenever `b` is not equal to zero, this line instructs the function to run again, this time with an incremented `a` and a decremented `b`. We thereby use `b` as an implicit counter telling us how many more times we have to increment `a`.

Again, no small change could have gotten us from the previous implementation to this one. That required additional creativity. This version is also hard to vary, except for being commutative. We could move things around a bit. There may be a function that lets us abstract out much of this. If you can change part of a well-adapted function *without* reducing its ability to solve the problem you designed it to solve, that is an indication of the reach of the replacement.

However, compared to the first implementation, I find the second one even harder to vary. The first implementation is adapted to the purposes of both looping and addition without being very well adapted to either one and as a result, it has some external variability. The second implementation does not loop. As a result, we require fewer variables and fewer lines, which makes the implementation easier to read and test. That is a special case of being adapted to a particular purpose; our previous function implementation was adapted to the purpose of addition and looping, but we were only after an adaptation to addition. So, simplifying functions and improving their structure is a good idea because it reduces internal variability, and often external variability as well.

The second function implementation depends on the concepts of `zero?`, `inc`, and `dec`, so it would be challenging to build without having those functions first. It also means that creative recombination of the concepts `zero?`, `inc`, and `dec` can lead to the entirely new idea of addition. Following Deutsch (2012a, 367), but intending a slightly different meaning, a function's underlying functions I consider *inexplicit*; to make them *explicit*, we have to look up their implementation. Inexplicit knowledge, says Deutsch, cannot quite be put into words, though in some cases we may be able to do so upon further inspection: I cannot tell you how I tie my shoelaces – I just *do* it – but if I watch myself closely as I do it, I can explain to you *how* I do it.

We want an implementation to be hard to vary while still accounting for what it purports to account for. Without that constraint, it is trivially easy to vary any implementation. There can be other constraints, too: we want our implementation to be hard to vary while using little memory, or while being fast. The more an implementation is constrained, the harder it becomes to vary. Sometimes individual function implementations or even entire programs can be so hard to vary that it is difficult to make changes and perform maintenance as a project matures. That is an unfortunate side effect of those implementations' not being adapted to the purpose of facilitating changes. The implementation resists change so much that it can cost a lot of time and money to break the resistance. One solution to this problem is modularity: programs should be hard to vary while still being modular. It is vital to make sure that implementations do not resist good kinds of change.

Here is an example of an implementation of that same function that is easier to vary internally:

```
(defn plus' [a b]
  (if (zero? b)
    (+ a 0) ; <— this change is useless
    (recur (inc a) (dec b))))
```

The change on line 3 is useless. It just adds `0` to the result. We can undo this change without reducing the whole function's ability to fulfill its purpose. Sometimes, removing internal variability can result in a good explanation, which has given rise to an insufficient concept called "Occam's razor." According to this concept, one should always prefer the simplest explanation. But simple alone often comes at the cost of external variability: "The gods did it" is a simple but highly externally variable explanation.

An example of a function with equally high external variability is:

```
(defn perform [f & args]
  (apply f args)) ; <— this line invokes
any function at all
```

All this function does is call any given function `f` with the provided arguments. It is easy to vary externally because it works for any function that is passed in, and so it is not adapted to a particular purpose; as a result, its structure does not explain anything. It just passes the job of explaining to the `apply` function, so one could just use `apply` to begin with. It is a kind of regress. It is a bit like watching a conjuring trick and explaining what happened by saying, "the conjurer did something." The explanation is not false, but it is bad. The "perform" function has tremendous reach because it applies to any function at all, but that reach comes at the cost of its adaptation to a purpose. It does not have a purpose, and this results in external variability.

We can combine both kinds of variability to get an example of a function that is easy to vary both internally and externally:

```
(defn perform [f & args]
  (+ 0) ; <— this line is useless; it adds
0 to nothing
  (apply f args))
```

It is easy to introduce internal variability by making almost random changes, which shows that easily variable functions are *not* discrete.

A real-life example of an explanation that is easy to vary internally is "you perform addition by repeated increments while wearing a hat." It is easily variable internally because you can improve the adaptation by removing "while wearing a hat." An example of an explanation that is easy to vary externally is "to do x you have to understand it first." Its internal structure is sound, consistent, and simple, but it applies to anything at all.

An example of an explanation that is both easily variable internally as well as externally is "The gods did it." It is easily variable internally because it is easy to replace "gods" with "angels" or "wizards," and it would still explain as much. And it is easily variable externally because it explains anything at all. I guess that most explanations that are easy to vary externally are also – at least to a degree – easily variable internally.

Here is another parallel between explanations and functions. Neither can ever be proven true, meaning it is impossible to test them exhaustively. Both are always only *tentatively* held as being true and expected to contain mistakes. There is not any value in repeatedly finding confirming instances of a theory unless they are part of a test that has some chance of falsifying the theory – a *severe* test (Popper 1983, 14-15) – otherwise, confirming instances just tell us something we already know. If you go to an oracle seeking a revelation and it tells you something you already know, you will not consider that a revelation (Deutsch 2012a, 225). Yet this is precisely what those seeking justified true belief claim.

For example, general relativity cannot be proven true because it is stated in such a way that it *is* true at all times everywhere. That is by design so that a single problematic observation could, in principle, refute it. In order to prove it true, you would need to fail to prove it false it at every space-time coordinate in the universe (Popper 2005, 47). That is impossible to do. The same holds for function implementations. No matter how many tests succeed, we have always only run a finite number of tests compared to the infinity of tests the function would need to pass to establish its truth, so we cannot assign probabilities either. As a result, functions, just like explanations, have a probability of being true of 0. It is just not a property that is of interest, and, therefore, we need not bother assigning probabilities to theories (Popper 1983, 265).

When we do assign a probability to a theory, that is a property of *us*: how much we are willing to believe that it is true. But it is independent of the theory itself, and attributing probabilities to our theories/functions would be a strange form of anthropomorphism. It is a *subjective* notion of probability, not an objective one – it confuses objective probability with a subjective experience of certainty (Popper 1983, 141).

A common yet divisive practice in software engineering is called "test-driven development" (TDD). It was popularized in part by modern web development frameworks and is equally loved by some developers as others despise it. It works by writing tests specifying the desired implementation before writing it. This way, failing tests act as a guide as to what remains to be programmed, which improves productivity.[5] Additionally, programs well covered by tests are harder to vary externally: when developers introduce changes later, tests may fail, alerting them of problems whose solutions require additional creativity. That boosts productivity once more since they can work without worrying about inadvertently introducing breaking changes: if they do so, failing tests help the implementation resist change by alerting developers about what they need to fix.

Of course, changes to implementations should also be reflected in tests first before making those changes. TDD requires some discipline to confer these benefits: as a project's complexity grows and ensuring external invariability becomes ever more critical, developers may nonetheless feel tempted to "just make a quick change" and write the tests covering that change afterward (or they conveniently forget to). This reversal reduces the role of tests to that of giving the developers certainty.

Contrary to appearances, this is not the purpose of TDD. Its purpose is error correction, which, combined with the developer's creativity, helps find a good implementation. While adding or updating tests after writing or changing an implementation still confers some, albeit poorer external invariability, the tests' primary purpose has become giving developers a false sense of certainty, which, adding nothing of value, eventually causes developers to give up the practice altogether. Abusing TDD in this way is another form of justificationism.

Overall, software engineering is wonderfully devoid of many philosophical misconceptions that have consumed some other fields. For example, there is no empiricism or induction in software engineering (at least when it is not applied to intelligence research, see chapter 7), because it is evident that the origin of programs cannot be observation, but only reason: the programmer's creativity. We cannot learn from sense data for the same reason one cannot leave a computer without any programs out in the sun (light as data) and wait for programs to write themselves on it.

However, the field is not immune to philosophical misconceptions. Even though the pursuit of certainty – justificationism – is not as common in software engineering as it still is in science and philosophy generally, it is not widely understood that "proving" that a function (or a scientific or philosophical theory) behaves as intended (or is true or whatever) is impossible. We require the understanding that all knowledge and all function implementations contain mistakes and grow only by the "method" of repeatedly and forever correcting errors.

Every software engineer knows this deep down: it would be ridiculous to show your colleagues a function you just built and pride yourself on how it has so far given the correct return values for every test to which you have submitted it. Instead, an implementation's foremost quality lies in its *structure* and its *reach*; that is, in its ability to solve a problem or even a large number of problems well. A programmer selects for this quality long before he begins to test it. He submits his functions to large amounts of criticism, not only before running them but mentally, *before writing them down.*

The very first tentative written implementation is the result of many iterations of conjecture and criticism in a programmer's mind. So we usually only perform tests after our criticism has improved a function's structure up to the (also conjectured) criterion of being sufficiently hard to vary while solving the problem at hand. When we *do* perform tests, it is always in the service of criticizing and improving the function, not to gain confidence that the implementation is correct and need not be improved further. It is only when a function implementation is hard to vary while still accounting for what it purports to account for that we even bother to run it or test it. As Deutsch says (2012a, 25), "Science would be impossible if it were not for the fact that the overwhelming majority of false theories can be rejected out of hand without any experiment, simply for being bad explanations." The same is true of software engineering: *it* would not be possible were it not for the fact that we can dismiss the vast majority of *implementations* out of hand simply for being bad implementations and without ever running them.

Some programmers may see this as an occasion for despair. They may say, "If I cannot become more confident that my programs are running correctly, why bother testing them at all?" That is easy: because the more errors you correct, the better your programs get. Say we get some part of the implementation of our addition function wrong, and when asked to calculate `4` plus `4`, our function does not return `8` but `22`. That means that part of our implementation is wrong, and with some effort, we can correct it. Once corrected, the function is objectively better adapted to its purpose – assuming that change did not introduce more issues.

Now, can we be *sure* that next time it calculates, say, `400` plus `271`, it does so correctly? Not until we try. Can we try every pair of numbers? No. Can we even be sure that the *same* pair of numbers will yield the same result the second time, and the third? No. We can have good explanations as answers to these questions, but we cannot be sure. If our function has a hard-to-vary implementation, that is all we are after, and we tentatively adopt our function implementation vis-à-vis the problem we designed it to solve.

That is fallibilism applied to software engineering: we should expect all our programs to contain mistakes in addition to working parts, and even mistaken programs often have tremendous reach and work quite well in their domain. There are tools available to programmers that generate test cases to catch flaws in programs, such as "QuickCheck." Programmers should use tools like it because they may help catch bugs. However, if they use these tools to become more "confident" in their code being bug-free, they are mistaken.

Consider this example of a program which appears immune to criticism:

```
(defn four []
  (* 2 2))
```

All this program does is return the number 4. It is the same example given by German humorist Wilhelm Busch that Karl Popper liked to quote as a "rhyme for the epistemological nursery" (Popper 1983, 54-55):

> Twice two equals four: 'tis true,
> But too empty, and too trite.
> What I look for is a clue
> To some matters not so light.

Popper addressed this by stating that truth is not the only aim of science. We want truth that is interesting and hard to come by; we want "solutions to interesting problems." (1983, 55)

It is not clear which problems the function above solves if any. But in any case, even this function is not immune to criticism. The implementation relies on multiplication to work, and implementing multiplication, like implementing addition, is indeed a matter not so light, and subject to all sorts of errors. Moreover, as I said earlier, we cannot know for sure that it will correctly return 4 every time we run the function. We can only guess, and that is enough.

What about an even simpler function?

```
(defn four []
  4)
```

It is not clear what problem this implementation solves, either. Perhaps it makes some other implementations that make use of it more idiomatic. The point, however, is that it only tentatively serves as an example of a seemingly secure implementation because of one's failure to criticize it. Failure to criticize is a reflection of one's abilities, not an indication that this implementation is immune to criticism. Tomorrow we may very well find something wrong with it.

The problem of induction has made an appearance in software engineering. For example, spreadsheet applications offer functionality that automatically completes a sequence for the user. Given a sequence of say, `1, 2, 3, 4`, the program automatically continues the list with the numbers `5, 6, 7, 8`, up to a useful limit. This functionality is not based on any alleged principle of induction, nor on the assumption that the future will resemble the past. The program that can detect and extrapolate such sequences is not "induced" from them: programmers evolved it in their minds beforehand. They simply *guessed* which one the user might mean. Knowing that their guess is as fallible as any, they let the user correct the suggestion.

Outside of software engineering, the cynical and widespread answer to the problem of induction is *relativism* (Deutsch 2012a, 15). It says that if we cannot be certain that our knowledge is true, or at least likely to be true, then we cannot say that one theory is better than another. Relativists conclude that all knowledge is on equal footing. This idea is a *meme* because it spreads from person to person, in part because it makes its holders appear to be particularly tolerant. It actively prevents problem detection and is the source of many bad ideas and complacency, including, ironically, tolerance for particularly brutal and intolerant ideologies.

For example, while the justificationist flavor of induction was a useful defense against religion – "if we cannot prove the existence of god, I will not believe in him" – its refutation has led to agnosticism: "we can't, like, *dis*prove it either." But no proof is needed to take sides on a matter; one can do that with the help of good explanations.

According to relativism, there are no conflicts between ideas, and there is no need to take sides. This assertion makes it a recipe for standstill. Fortunately, there is no selection pressure for appearance of tolerance in software engineering: the bar for good function implementations is higher than it is for good ideas generally, and function implementations are always judged by how hard to vary they are, as well as by their reach. Software engineers embrace the notion that objective differences in quality between implementations not only exist but provide the possibility for improvement. If they did not, the whole enterprise would die because they could not discover problems.

Just as a bad explanation can still explain something and may even be true, a bad function implementation does not necessarily mean one cannot run it, and it may even solve one's problem. However, when we *do* find an issue with an explanation or a function implementation, we want to avoid so-called "ad-hoc fixes." An ad-hoc fix is a fix whose introduction does not require much creativity and does not explain why the function failed in the first place. To give the same example a second time, what if our addition function worked great, except for some reason it returned 22 instead of 8 as an answer to 4 plus 4? Consider this tentative solution to the problem:

```
(defn plus-ad-hoc [a b]
  (if (and (= a 4) (= b 4))
    8 ; <— hardcoded answer for 4 plus 4
    ; ... faulty addition code here
  ))
```

Before performing our faulty addition, we check if the parameters are both 4 since we know they will lead to an error. We end up bending over backward to make this thing work by hard coding the solution for that case, rather than solving the underlying addition problem. We "immunize" (Popper 1983, 30) our function against this kind of criticism.

That applies to explanations as well. In 1919, an experiment was conducted during a solar eclipse to test Einstein's then-newly proposed general theory of relativity; it ended up corroborating his theory. Imagine if that experiment had gone differently and had ended up refuting the theory. If he had then said, "well, the eclipse fell on a Thursday, and general relativity works only on all other weekdays," not only would he have immunized his theory ad hoc – he would also have lost all credibility as a scientist. By the way, the experiment on that day never "proved Einstein true," as many claim. It was merely a crucial test that failed to refute the theory.

The similarity between epistemology and the rules of how to write good computer programs does not stop there. We can even translate some philosophical misconceptions into function implementations that work poorly. Consider the doctrine of instrumentalism once more, which states that science cannot describe reality and that therefore one should not even bother explaining anything and only make predictions. Applied to software engineering, it is the doctrine that programs are only useful and do not explain reality. But there is no such thing as an "explanationless prediction" (Deutsch 2012a, 15) for the same reason there cannot be a return value without a function. Moreover, if an explanation is useful, it is always because it contains some truth. So even instrumentalist interpretations of software engineering rely on programs containing some truth because it is this truth that makes the program applicable and useful in the first place. It is this truth that allows us to exert some control over the physical or abstract world and, therefore, to consider the program useful in the first place.

The following is an example of a poor explanation of addition that resembles instrumentalism:

```
(defn plus-instrumentalist [a b]
  (cond
    (and (= a 1) (= b 1))
    2
    (and (= a 1) (= b 2))
    3
    ; ...
    (and (= a 2) (= b 1))
    3
    (and (= a 2) (= b 2))
    4
    ; ... and so on up to some limit
    ))
```

This reads, "If `a` is `1` and `b` is `1`, return `2`; if `a` is `1` and `b` is `2`, return `3`," and continues to do so up to some limit. The possible answers for some pairs of `a` and `b` are hardcoded.[6] This function is such a bad explanation of how addition works that no one would ever write such a program. Nevertheless, this is precisely what instrumentalism claims science is about! This function also shows why explaining addition well is better than just giving someone an addition table: such a table has minimal reach as it always only extends up to some predefined limit depending on how many results were pre-programmed. It is also always easy to vary internally: we can easily replace each range of covered integers by a different range of integers without making the program better or worse.

Due to its poor structure and limited reach, the implementation can be considered *non-explanatory* for addition – but it is nonetheless explanatory for mapping specific pairs of numbers to results. It is not entirely explanationless either, for it relies on underlying explanations of mappings, equality, and others.

According to Deutsch, *behaviorism* is just instrumentalism applied to psychology. It is the doctrine that science can (or should) only explain behavior as the result of conditioning. This doctrine has damaged AGI research in the form of the Turing test and reinforcement "learning," which I will examine in chapter 7.[7]

Some bad explanations involve something called an "infinite regress." For example, the claim that design requires a designer then begs the question "well, who designed the designer?" and so forth, ad infinitum. Stephen Hawking tells the following story (1998). (There are variations of the story, but his account is as follows:)

> A well-known scientist (some say it was Bertrand Russell) once gave a public lecture on astronomy. He described how the earth orbits around the sun and how the sun, in turn, orbits around the center of a vast collection of stars called our galaxy. At the end of the lecture, a little old lady at the back of the room got up and said: 'What you have told us is rubbish. The world is really a flat

> plate supported on the back of a giant tortoise.' The scientist gave a superior smile before replying, 'What is the tortoise standing on?' 'You're very clever, young man, very clever,' said the old lady. 'But it's turtles all the way down!'

Of course, explaining the world this way is a bad explanation because it does not explain anything. It forever passes the explanation down to the next turtle. This problem can be found in computer programs as well in the following form:

```
(defn tortoise []
  (tortoise))

(defn earth []
  (tortoise))
```

You may choose to read this from bottom to top. When you try to invoke the `earth` function, it will, in turn, invoke the `tortoise` function, which, from then on, will keep invoking itself. It will keep trying until the computer runs out of memory to keep track of the progress it has made in calling the next function. This phenomenon is known as a "stack overflow." However, the deeper reason this is problematic is not just limited memory, but that the `tortoise` function does not explain anything. It just keeps delegating – incidentally, to itself.[8]

All of these examples of laws of epistemology applying to computer programs serve a particular purpose, namely, the purpose of suggesting an answer to the question of how we can encode explanations in computer programs. Let us revisit the good implementation of addition from earlier:

```
(defn plus' [a b]
  (if (zero? b)
    a
    (recur (inc a) (dec b))))
```

As I was sitting in that classroom when I was seven or eight years old, my explanation of how to perform addition looked something like that. Presumably, so did yours. I have rarely run this function since then because I have memorized the results for small integers. Likewise, the results of certain function invocations in computer programs can be cached, or "memoized" (sic) – "remembered," in a way. For larger numbers, as well as for decimal numbers and fractions, I use different algorithms. But all this is to say the similarities between the laws governing what makes a good explanation and a good function are not accidental. I guess that that is so because this implementation *is* a hard-to-vary explanation of addition. Therefore, I guess that *function implementations are explanations*.

This equivalence will make several appearances throughout this book. Functions are knowledge encoded in computers. A function's body *is* the explanation's body; the function's parameters are the experimental setup or the initial conditions; the return values are the function's predictions or claims; running a function is to make a prediction, and running two or more functions expecting specific return values is to test them in order to detect problems. Since functions and explanations share all of these properties, explaining explanations in terms of functions is itself hard to vary. That gives us another way of explaining explanations: they are computations that are adapted to solving problems.

My teacher knew how to perform addition. He had a function in his mind that he had initially created to solve the problem of how to do so. He wanted to pass this knowledge on to us, the students, and so we had to recreate it. By recreating it, we understood it. The reason we do not think in terms of computer programs from day to day is that explaining things is second nature to us.

Notwithstanding, I conjecture that every explanation we as individuals have is a program that runs in our minds, which we can invoke to make predictions or to test it. The origin of each function is our creativity, which means the creative algorithm returns functions. This process works not just for functions in terms of mathematics, but for *any* kind of function: understanding and giving directions (Deutsch 2012a, 254), understanding a recipe from a cookbook, how to drive a car, or play the piano... Anything at all that we learn we store in our brains as a function implementation. To understand anything means to write a program of how it works in one's mind.

When there is a regularity in nature, we seek to explain it. By doing so, our explanation not only incorporates that regularity, it *is* a functional representation of that regularity in our minds. It replicates the regularity. Therefore, replicating a function is the computational counterpart to creating an explanation. Due to the equivalence between functions and explanations, and the laws of epistemology applying to both, I use the terms "function," "program," "explanation," "theory," and "idea" interchangeably throughout this book. They are all the same.

Terminology

Function – Explanation, that is, a computation adapted to solving a problem

Function implementation/body – The explanation's structure

Function parameters – The experimental setup/initial conditions

Return values – Predictions/claims

Problem – Conflict between two or more functions

Explicit/inexplicit knowledge – A function whose implementation is currently known/unknown (can/cannot be easily put into words)

Meme – A function that spreads between people

Behaviorism – Instrumentalism applied to psychology. The doctrine that science can (or should) only explain behavior as the result of conditioning

To explain something – To create a function, if only mentally, that solves a given problem

Summary

AGI creates explanatory knowledge by definition. If we want to build it, we have to find a way to encode explanations in programs. The solution is that explanations *are* those programs: explanations are functions and vice versa. They are identical. That is the reason the laws of epistemology apply to computer programs.

Functions are autonomous, and their content is independent of their creator. Good function implementations are discrete, meaning a "valley of bad function implementations" has to be traversed in order to get from one good implementation to another vis-à-vis the same problem.

Function implementations solve problems. They are created through creativity, not mechanistically. Being able to replace part of a well-adapted function without making it worse is an indication of the reach of the replacement. Well adapted function implementations resist change.

No function can ever be proven true, or likely to be true; instead, we hold functions to be *tentatively* true. We test programs not to gain confidence in their running correctly, but to catch and correct errors.

A function's quality lies in its implementation's structure and reach. Some epistemological misconceptions have manifested, albeit less obviously, in software engineering, e.g., ad-hoc fixes, instrumentalism, and infinite regresses.

Knowing how to encode explanations in computer programs, we can, in principle, build a universal explainer: an AGI. But what is universality?

4
Universality

"Man's reach exceeds his imagination"
– *Movie "The Prestige", 2006*

Reach is a desirable quality of explanations: an explanation with greater reach can explain more than one with little reach. Every explanation has at least *some* reach: at least enough to solve the particular problem it purports to solve, however poorly it may do so. The greater an explanation's reach, the more problems it solves within a domain, or even across domains, even though people may originally only have designed it to solve one specific problem. When it solves all problems in a single domain – or at least can do so – it has *universal* reach within that domain. That is *universality*. I will explain the main concepts related to universality based on Deutsch's work on it (2012a, chapter 6) and some elaborations by Temple (2019d and personal communication with author, 2019).

The concept of universality not only applies to explanations, but to anything that works within particular, arbitrarily defined domains. Consider the following example: a car is a universal car if it can drive on all roads that all other cars can drive on. It does *not* necessarily mean that a universal car can drive on *any* road. Instead, it is a measure of an ability's reach within a domain: how many of the problems in that domain can the ability solve? If all of them, it is universal (Temple 2019d). Some cars are designed to drive on city streets, while others are designed to drive on race tracks, and others still are designed to drive on muddy roads. If we think of a domain of "general-purpose driving" that consists of driving on city streets, race tracks, and muddy roads, a car that is universal in that domain can drive on all three.

Universality is a binary matter (Deutsch 2012a, 157). Something either is universal in a domain, or it is not. In finite domains, it may be useful to speak of something being "almost" universal. For example, a car that can drive on city streets and race tracks, but not muddy roads, may be considered an "almost" universal car according to our definition. But in infinitely large domains, there can be no such thing: being able to do, say, 30 out of infinitely many things is still precisely 0%.

We may choose to divide domains into sub-domains (Temple 2019d). For example, we may consider a car a universal *race* car if it can drive on all tracks all other cars can. It does not mean a universal race car can drive on *all* tracks unless we change the domain's definition to mean that (Temple, personal communication with author, 2019). We performed this kind of subdivision of domains in chapter 3 when we discovered that our addition algorithm is universal for *positive* integers: it can add all positive integers any other such algorithm could. But it is not a universal addition algorithm for all integers, including negative ones, because it does not explain how to perform addition when the second parameter is negative, which any other such algorithm would explain.

Universality has played a role in our evolution and essential inventions throughout history. An example of a universal invention is one that occurred in printing technology. Up until the invention of the movable-type printing press, people had developed different ways to print books. They either copied books by hand – a laborious, expensive, and error-prone task – or carved the mirror script of each page onto a dedicated printing plate, which was then inked and pressed onto a new piece of paper. It is akin to having a rubber stamp for each page and using a collection of those pages to print the same book several times. This process was less laborious and less expensive, and it also had a chance of being less error-prone because mistakes such as misspellings could only happen once during transcription onto plates. After that, while one could repeat the same mistakes, one could not introduce new ones by the same means.

While this approach was better than doing it by hand, it had limited reach: a stack of printing plates could be used to print only the book it represented. Movable type changed this. While many accounts of the history of movable type focus on improvements related to speed and efficiency, it made for a *qualitative* improvement as well. Rather than carving out each page, one simply arranged movable letters to represent the page one was copying. Instead of creating customized printing plates, one could reuse movable letters, that is, customize the universal system. I say that movable type turned the printing press into a *universal* system because people could use the same type to print *any* book that printing plates could print.

Printing plates only shared the repertoire of supporting the print of a particular book. A printing press powered by printing plates encoding the bible, for example, could be described as a "universal bible printer": it could print all the books all the other bible printers could print, namely the bible. But that is not a particularly interesting kind of universality. A printing press powered by movable type could print any book whatsoever, and therefore automatically supported the repertoire of all presses powered by printing plates, as well as all other movable-type presses. It was a general-purpose printing press. It even supported the printing of books that had not been written yet! So movable type was a *jump* (Deutsch 2012a, 146) to two kinds of universality at once: first, it was universal in the domain of printing books that all existing printing plates could print; second, it was universal in the domain of printing any book at all – regardless if it had been carved into printing plates, or if it had even been written yet. The latter domain happens to include the former. That is an example of the momentous reach of movable type in the domain of printing books. This domain is infinitely big: there are infinitely many books one can write and then print.

In infinitely large domains, universal explanations have infinite reach. Such an explanation does not necessarily apply to *everything*; it has total reach in an infinitely large domain. That domain can still be exceptional or limited (Temple 2019d). Universalities in infinitely large domains can only be achieved in a jump because making single additions to a list of features never makes that list infinitely long. For example, imagine yourself carving out printing plates for the bible. Then you carve out plates corresponding to "Romeo and Juliet." You can keep doing that for as many books as you like; it will not get you any closer to a printing press that can print any book, because there are infinitely many possible books. Once there is a new book that needs printing, you cannot print it until you create new printing plates dedicated to that book; unless you have movable type or some other general-purpose printing technology.

Whenever something reaches universality in a given domain, it is necessarily the same qualitatively as everything else with universality in the same domain. The repertoires are the same. Why? Because all entities with universal reach in a domain necessarily have the same reach in that domain. For example, all movable-type printing presses can print the same books. These printing presses can differ in how many movable letters they have, or how fast they allow you to print; but in terms of functionality, they are all the same. Universal reach in a domain is the maximum reach that can be acquired in that domain; by definition, there can be no reach beyond universality.

Even printing presses' constructions will not matter as long as they support the same repertoire because there is a difference between functionality and *hardware*. As long as the hardware supports the universal repertoire – in this case, the printing press's material – it does not matter what the hardware is. The universal repertoire transcends its hardware. That is always the case for hardware whose functionality has infinite reach in a domain, while its physical extent can always only be finite. Its functionality reaches beyond its hardware limitations into infinity.

Computers have reached universality in a particularly exciting domain. A computer being universal per se does not tell you what that computer can do: it just means it has the same functionalities as all other computers.[9] But what computers *can* do is run any computable algorithm, and by doing that, they can simulate any computable physical process and computable abstraction with arbitrary precision (Nielsen 2004). The domain of possible physical processes is infinitely big, and all general-purpose computers have universal reach within it. They are all qualitatively the same and can only differ in terms of memory and processing speed. All general-purpose computers share the same repertoire, from our present-day laptops, desktop computers, and smartphones, all the way back to the first general-purpose computers from the 1940s, as well as any computer that we *could ever build* in the future (Deutsch 2015, 48:47).

This kind of universality is called "Turing completeness" (named after Alan Turing): a computer is Turing complete if it can simulate any Turing machine. Whenever I refer to computational universality, I mean both the shared repertoire of computers *and* their ability to simulate any computable physical process and abstraction with arbitrary precision.

Most modern programming languages are approximately Turing complete.[10] A Turing machine is an abstract machine that one can think of as containing and operating on an endless tape divided into cells containing, for example, digits 0 and 1 (bits). These bits may represent your program and/or data you are working on. There is a read and write head that scans each cell and changes its value to either 0 or 1 or skips it, and then moves to the next or previous cell, all based on the cell's value and arbitrary instructions given to the head. It keeps doing this until it encounters an instruction to halt. This halting instruction may never come, but if it does, you can then look at the tape and find the answer to whatever you are trying to compute. This ability alone is powerful enough to compute anything that can be computed.

Before computational universality, people had to build dedicated machines for each problem. That is analogous to printing plates for printing books. Today, we have general-purpose computers, which need only be customized/programmed to solve a particular problem. Being customizable as opposed to dedicated is a characteristic of universal systems. Customization of hardware is known as *software*, which can be stored in the hardware itself, or the minds of *people*. So far, software has mostly been created genetically and in the minds of people; figuring out a way to build a program that creates software on a computer is part of the problem of how to build *AGI*.

There are different approaches to Turing completeness. Babbage's Analytical Engine would have been the first Turing-complete machine had it been built at the time it was designed. Alan Turing discovered Turing machines in an attempt to solve the mathematician David Hilbert's *Entscheidungsproblem* ("decision problem"). It can simply be stated as "is there an algorithm that can decide every problem in mathematics?". (The answer is "no," but the reason is out of scope for this chapter.) To solve the decision problem, another problem had to be solved first: one had to formally define what an *algorithm* is. Up until that time, the answer had informally been that anything a mathematician could calculate with paper and pencil was an algorithm. Turing's more formal solution was: anything that can be computed by a Turing machine is an algorithm.

A Turing machine is an abstract, special-purpose machine designed to run a particular algorithm. But *universal* Turing machines can simulate any Turing machine. Around the same time, the mathematician Alonzo Church, Turing's Ph.D. supervisor, independently came up with a different answer to this question: any function that can be expressed in *Lambda Calculus* is an algorithm. Turing immediately understood that both abstract concepts of Turing machines and Lambda Calculus were equivalent models of computation: they can be used to simulate each other. Lambda Calculus is a remarkably simple concept. It relies only on three simple rules:

1. One can declare variables, such as `x`, `y`, and `z`
2. One can declare single parameter functions, such as `(fn x . x + 1)`: This is a function that takes a variable x and increments it by `1`. (I have chosen to alter the standard syntax slightly because it is confusing.)
3. One can invoke a function: `((fn x . x + 1) 5)` This example will evaluate to `6`.

The details of these rules are not all that important for our purposes; suffice it to say that, although there are only three of them, and despite their simplicity, they are enough to make Lambda Calculus computationally universal. That means, in turn, that the functional approach discussed in chapter 3 has universal reach as well, which it needs to have in order to power general-purpose creativity. Both Turing's and Church's solutions to the question "What is an algorithm?" boil down to this: anything that can be computed by a machine is an algorithm. This is known as the Church-Turing thesis. By solving this problem, both Turing and Church designed computationally universal systems.

Let us look at an example of a "jump to universality" (Deutsch 2012a, 146) in code. Imagine programs dedicated to greeting a person:

```
(defn greet-tom []
  (println "Hello, Tom"))

(defn greet-barbara []
  (println "Hello, Barbara"))

(defn greet-marcus []
  (println "Hello, Marcus"))
```

Each of these programs merely prints a greeting to the screen. They all have the dedicated purpose of greeting a particular person. What follows is a program which, through a minor tweak, has been allowed to jump to universality in the domain of general-purpose greeting. It can greet anyone all the other greeter programs can, indeed, that all other *possible* greeter programs can:

```
(defn greet-person [name]
  (println "Hello, " name))
```

Instead of greeting a particular, hardcoded name, the program is customizable. It takes the name it is supposed to greet as a variable. To a programmer, this may seem unremarkable, and it is the obvious solution to the problem of having repetitive code of this kind. But such a solution can always only seem unremarkable in hindsight, once one has created the requisite knowledge to execute the jump to universality. And repetitive code can only be recognized as a problem once one knows that one should avoid such repetition. None of us are born with that knowledge. It is why I guess that if you are not a programmer, repetition does not seem problematic. In any case, the idea of having variables is hard to vary externally while increasing a program's reusability *despite* its tremendous reach.

When a genetic mutation introduces a universal program in *genetic* code, its special-purpose counterparts will remain present for several generations at first. But mutations that degrade them will not be selected against since the universal program has replaced their functionality. Eventually, the special-purpose programs will fade away over the generations.

It takes additional effort to build every additional greeter program as well as maintain all of them. A general-purpose greeter is easier to build. If you need to greet someone else, you have already done all the work to do so. Avoiding universality makes the problem *harder*. It is for this same reason that we use general-purpose computer chips for dedicated applications and in special-purpose machines such as digital thermometers and thermostats.

This universal greeter reaches even beyond general-purpose greeting. It allows you to greet someone whose name is not known at the time of writing the program, but only at runtime, perhaps because the user typed in his name. That is an example of how, once a jump to universality has been achieved, the universal system enables you to work on new problems that you did not consider soluble before. It is this reach of some generic solutions to specific problems that allows biological evolution to work despite often getting stuck in local optima: sometimes, evolution stumbles upon a generic solution to a problem and enjoys the advantages of its reach. Likewise, even though, as a child, you created the knowledge to perform addition primarily for application within a small range of integers, you get an algorithm that can do it for any integer for free: the implementation has universal reach.

Devices with computational universality are remarkable because, as I have said, they can simulate every computable physical process. This is known as the Church-Turing-Deutsch principle, named after Alonzo Church, Alan Turing, and David Deutsch, each of whom contributed to it. If you can find it out there, in the universe, you can simulate it on a computer. This includes *emergent* properties of people, such as consciousness, emotions, humor, and creativity. (Emergent phenomena are those that are best explained without referring to their lower-level components, such as hardware (Deutsch 2012a, 107-108).) I choose these properties because they are held by many as being utterly mysterious and uniquely human, and often mixed with the romantic belief that they cannot possibly be simulated. It is a special case of the general belief that some things "simply cannot be explained," sometimes coupled with the anti-rational notion that if everything were explicable, something would be "lost." But this is a bad explanation because it amounts to a belief in the supernatural, and it is a false explanation because it violates the Church-Turing-Deutsch principle.

The reach of general-purpose computers still pales compared to the reach of *people*. It is not that people can simulate things computers cannot – that would violate computational universality – it is that they can *create* those simulations, too. In contrast, computers currently still need to be fed the requisite software by a person. At a specific point in their evolution, on which point I will elaborate in chapter 5, people reached universality in the domain of explaining: they are *universal explainers* (Deutsch 2012a, 146).[11] People can create new knowledge to solve any soluble problem, of which there are infinitely many. There can be no greater reach than this. All people have the same ability to create new knowledge at runtime, and can only differ in terms of memory, processing power, and the knowledge they contain. (In this case, by "runtime," I mean a person's lifetime.)

There are no guarantees: they will make mistakes, and they may choose not to solve problems. But if they do, they *can* succeed. The solution to any problem is within their reach. Since people can, in principle, create any possible knowledge, the reach of people is the same as that of all possible knowledge, and there is nothing that can be done that cannot be done by people (Deutsch 2012a, chapters 3 and 9).

Most ancient accounts of reality were anthropocentric, and a departure of anthropocentrism has characterized much of scientific progress since. We are not located at the center of the solar system, nor of the universe, nor was the world made for us by a god. Nonetheless, people *are* significant in the cosmic scheme of things (Deutsch 2012a, chapter 3) because of their reach. Indeed, they are the *most* significant entities in the universe because nothing can surpass their reach.

AGIs, by definition, are universal explainers as well. As I said above, systems that reach universality in a domain are qualitatively the same. It follows that AGIs are literally people. This fact addresses moral concerns about AGI, such as potential dangers – more on that in chapter 8 – and how to treat it.

It is not necessary to have a human body in order to create knowledge. Physical differences are irrelevant as long as they don't impede a person's creative abilities. The usual list of physical insignificances includes skin color and sex. We should add to this list any physical features whatsoever: it does not matter if a person is made of flesh and bones or metal and silicon; if his body is that of a human being or that of a desktop computer. You could run an AGI in a *web browser*, and it would be a person just the same. They are all people as long as they have a creative mind. The only physical feature that is required is that of providing universal computation. Distinguishing between carbon-based and silicon-based people is a new kind of racism (Deutsch 2015, 1:00:14). A person is a running instance of the creative program (Deutsch 2016b). Writing the creative program is its conception; running it is its inception; removing part or all of its memory without its consent is lobotomy, and terminating it without its consent is murder.[12]

Our ideas about what it means to be human are still affected by romantic theories: that it is our ability to love, hate, dream, and cry that makes us human. All of these result *from* this single, incredible ability people have – creativity – but none of them is the defining characteristic.

AGI research is a *software* project; it is not about building robots, or any other kind of hardware. We already have the requisite hardware: Turing-complete computers. General-purpose creativity requires Turing-complete computation because otherwise, there would be a limit to which functions people can guess, that is, which theories they can guess, and run. But there appears to be no such limit for humans. People have successfully guessed explanations allowing them to design and engineer space shuttles, medicine, bridges, and computers. The human brain cannot be a special-purpose machine because its mind can, in principle, guess the implementation of any function and then instruct the brain to run it. Since humans are creative, their brains must be Turing-complete computers.

That is not an analogy: it is literally so, and it follows directly from our best explanations of creativity and universality that anything that has the reach to simulate general-purpose creativity *is* a *universal computer*. Any serious critique of the idea that the brain is a computer would have to refute the universality of computation successfully. That would involve finding a soluble problem whose solution can be computed by a machine that is not a Turing machine – allegedly, for example, the brain – but not by a Turing machine. The critique would also need to explain how to answer the following questions in the negative: when someone solves a problem, does he not *compute* the answer? When someone calculates `4 + 3`, or walks to a store, or reads a book, or does anything at all, does he not run a particular program in his mind which allows him to do so? Is this program not software, and does software not run on a computer? Is it not information processing, and is all information processing not done by computers?

In any case, according to our best explanation of people, they are universal explainers (Deutsch 2012a, 157). We know of no other animal that is a universal explainer – although it is possible many animals have Turing-complete brains and could, in principle, be creative if they possessed the creative algorithm. We know of no extraterrestrials who are creative, either. Nor are our computers – yet. Once they are creative, they are people just the same.

Terminology

Universal computer – A universal Turing machine. Given enough time and memory, it can simulate any computable physical process and abstraction with arbitrary precision

"Jump to universality" – A sudden increase in reach from limited to universal (Deutsch 2012a, 146)

Lambda Calculus – A Turing-complete, functional alternative to Turing machines invented by Alonzo Church

Hardware – A machine's raw materials

Software – Instructions customizing hardware

Person – A universal explainer/problem solver; a running instance of the creative program (Deutsch 2012a, 414)

AGI – A person running on a computer other than the brain

Universal explainer – A program that can solve any soluble problem; a person (Deutsch 2012a, 146)

Soluble problem – A conflict between ideas for which problem there exists a possible function that is a solution

Runtime – The time between the launching of a program and its termination. For people: their lifetime

Emergent – Phenomena are emergent when they are best explained without referring to their lower-level components (Deutsch 2012a, 107-108)

Summary

Universality is total reach in a domain. In infinitely large domains, universality can only be achieved in a jump, not gradually. All things that are universal in a given domain necessarily have the same repertoire. Beyond its influence on reach, hardware specifics do not matter for this purpose.

Functions are Turing complete.

Genetic mutations can introduce a jump to some universality, which explains the reach of some genetic adaptations. A mutation with universal reach may be able to spread more quickly through the population than its non-universal rivals.

Universal function implementations are often simpler and easier to build than non-universal ones.

People have the greatest possible reach: they can, in principle, solve any soluble problem because they are universal explainers.

We now understand that creating an AGI is a software project that involves building a universal explainer. Now, what is the logic behind explaining? How does creativity work, in detail, and how might one use that knowledge to build AGI?

5
Creativity

Creativity is the capacity to create new explanations. It is synonymous with *intelligence*. It is the algorithm that allows people to *learn*, that is, to create new knowledge by solving problems. It is what turns an otherwise automated organism into a person; it makes people, people. It is the source of all human knowledge, from the lowest level of abstraction to the highest, and from the lowest level mechanisms of learning to the higher levels, adding up to an emergent "self," an "I." It consists of the single activity of solving problems.

As I discussed in chapter 2, how creativity works is widely misunderstood. Empiricism, induction, Lamarckism, the psychology of learning, and other misconceptions all explain learning in terms of following some predefined method that leads to success. In reality, there is only trial and error correction. The whole point of creativity is to create *new* ideas and methods (what I have been referring to as functions), rather than merely following pre-existing ones.

Another widespread misconception is that we create different kinds of knowledge through different means. But, as I discussed in chapter 3, the logic by which children learn is the same logic by which scientists create new theories. Popper realized this early on while teaching, even though he was still an inductivist at the time (Deutsch, personal communication with author, 2019). This realization is much more profound than meets the eye: it means that we copy knowledge by the same algorithm by which we create it (Deutsch 2012a, 411). So, if you write a program that can *copy* any existing knowledge, you will automatically have built something that can create new knowledge, too. This approach may be viable to building AGI: it may well be easier to focus on building a program that *copies* knowledge than one that creates *new* knowledge. If your program can copy a function's implementation without reading it, based solely on inputs and outputs it tries out and perhaps some clues about its implementation, you may be on your way to building AGI. However, care should be taken for this not to devolve into a pattern recognition algorithm, which would not be creative at all.

In any case, whatever knowledge children create in, say, English class, they create through the same means that create scientific knowledge. In other words, knowledge about, in this case, the English language is created through the same means that create knowledge about the physical world (scientific knowledge). That is because creativity is *universal* in the domain of creating general-purpose knowledge. It can produce scientific, mathematical, philosophical, aesthetic, practical, moral, emotional, and technological knowledge, and whatever other types of knowledge there may be that we have not yet determined.

I say "determined" rather than "discovered" because, as Popper pointed out (n.d.), "subjects" do not exist. There are no dividing lines between these fields; people, Popper said, draw them as a matter of convenience, and so that schools can appoint experts and test students. But a student of anything quickly finds that all these subjects are related and interconnected in fascinating ways and that the limits of even our best understanding are usually just one or two questions away. Therefore, it is more productive to speak of problems rather than subjects.

Popper stresses that of all subjects, the scientific method exists the least: there is no definitive method one can follow to create knowledge. It is all trial and error correction. With some luck, a scientist *may* find a solution, but there is no reliable way to do so – and, according to Popper, many scientists find only one or two solutions to big problems in their lifetime, and then die before finding another. If there were a way to create knowledge *reliably*, says Popper, they would solve many more problems than they do. It is not like great scientists make an important discovery and then forget how the scientific method works. There is no such method. Instead, says Popper, scientists (and, I should add, all people) make their discoveries *through happy accidents*.

Creativity creates kinds of knowledge which may, at first, seem different in some way, such as *moral* knowledge. In the empiricist worldview, how we come by moral knowledge is utterly mysterious.

A variant of this problem, going back to the philosopher David Hume, is usually stated as "we cannot derive an ought from an is." At most, empiricism can claim to explain how we create *scientific* knowledge in particular because at least science is that part of philosophy that one can test empirically. Observations play *some* role in science, especially that of deciding between guessed theories (Deutsch 2012a, 32). But as I have explained, empiricism is entirely false, so the problem "we cannot derive an ought from an is," like the problem of induction, is a pseudo-problem. It is rooted in the misconception that we *derive* knowledge from something when really we *create it from nothing*.

A piece of moral knowledge, in particular, is a guess we create in response to the problem of what to do next (Deutsch 2012a, 121). There is nothing mysterious about it. So why should the creation of moral knowledge not follow the same logic as the creation of scientific knowledge, or, more generally speaking, why should the creation of one kind of knowledge not follow the same logic as the creation of all other kinds?

Following Deutsch, I believe that emotions are also just a kind of knowledge (personal communication with author, 2020). The unity of knowledge extends to more things not traditionally considered knowledge, including hopes, goals, and desires. For the particular case of preferences, consider this simple implementation of a preference function:

```
(defn determine-preference
  [criterion-fn & choices]
  (->>
    choices
    (sort-by criterion-fn)
    first))
```

We are, as always, not all that concerned with the specifics of the syntax, but I will give a brief overview. The function is given both another function that represents a criterion and an arbitrary number of choices. It then sorts the choices based on the criterion and returns the first, that is, the best one. The criterion function is itself conjectured based on context. For example, it may return values based on how sugary a given donut is, in which case the choices passed into the parent function consist of a collection of donuts, and the `criterion-fn` returns a number based on sugar content. If you like sugary donuts, your preference function will return the most sugary.

Again, do not be deterred if the syntax seems overwhelming. What matters is, this function is adapted to the purpose of solving the problem of determining a preference among choices. It is knowledge that can be created by a universal explainer. The fact that it is about forming preferences rather than a scientific matter makes no difference.

What about *qualia* (singular quale), the "subjective [aspects] of a sensation" (Deutsch 2012a, 163) such as seeing a color or feeling pleasure or pain? They are one of the deepest and most troubling problems today because so far we have been unable to describe or predict them (Deutsch 2012a, 153). While qualia are mysterious, that does not need to cloud our understanding of things like emotions and preferences being a kind of knowledge. Consider the above example again with a slight modification:

```
(defn determine-preference
  [criterion-fn & choices]
  (->>
    choices
    (sort criterion-fn)
    first
    (trigger-quale "determination")))
```

This implementation is the same as the original one, except for an additional invocation of a "quale of determination" at the end. This single step is admittedly mysterious – we have no good explanations of qualia yet – but all surrounding steps are explicitly known, so knowledge is not mysterious just because it invokes qualia.

Even actions are knowledge. They are functions that cause *side effects*. Side effects are changes in the behavior or state of the outside world, such as printing something on a screen, or making a web request, or even something as sophisticated as moving a robot arm. Indeed, that is how you and I move *our* arms, walk around, and effect physical changes in the world generally. A crucial difference between software written by the mind to, say, go the store, pick out a steak, go home and cook it as a solution to the problem of how to not be hungry, and a scientific theory such as general relativity, is not their complexity or subject area, but that the former has side effects while the latter does not. They are both functions: knowledge in the general-purpose sense. (And, since the mind can create functions that have side effects on its body, we can answer the mind-body problem – can the mind affect its body? – in the positive.)

The set of a function's side effects is its *phenotype*. Those phenotypes are what we see all around us: buildings, airplanes, conversations, art, computer games, and any other physical manifestations of knowledge. In most cities we observe a bunch of, say, cars, but a car of a particular make and model is not the replicator. It is the abstraction *behind* the car – the function – that is the replicator, and the car is part of the function's phenotype. The same is true for genes as well: it is not the DNA string that replicates (that would require cloning). It is the function behind the DNA string that replicates and instantiates many DNA strings encoding *it*. So not only do genes use organisms to spread through the population (Dawkins 2016); additionally, *functions use genes* in order to spread through the population.

We do not need to carefully examine different kinds of knowledge such as emotions, hopes, goals, desires, and preferences to conclude that they are all functions: we know this from the universality of computation. The brain is a computer, so *everything* it runs *must* be a function (that includes the ever so elusive qualia). Since creativity can create all of these kinds of knowledge, and since AGI, like all people, is an instance of the creative program, I conclude that an AGI can and will have goals, desires, preferences, emotions, and qualia as well.

As I explained in chapter 3, even a computer without the creative program knows things in the objective sense: all the knowledge that people have given it in the form of programs.

For example, it knows how to run a word processor, or it knows how to calculate a geographical route through traffic from point a to point b. But no computer can create *new* knowledge as of yet. That requires a creative program: an AGI.

Animals know things too: a dog knows how to, say, recognize things by scent and how to fetch a ball. This knowledge is given genetically. Knowledge is given to humans genetically as well: for example, the knowledge of how to detect some rotten foods by smell and how to heal a wound by growing new tissue. However, once again, people have the additional capacity to create *new* knowledge at runtime.

Books *encode* knowledge, but this is different in several ways from computers and living organisms: books contain symbols which, given the requisite knowledge, can be turned into the knowledge they represent. Popper instead claimed that books *literally contain* knowledge (1983, 107). However, books do not *contain* knowledge in the sense that they contain functions, let alone run them. Therefore, we can think of a hierarchy of things in terms of *encoding* knowledge (books), *containing* knowledge (animals, computers), and *creating* knowledge (people; they create *and* contain knowledge).

Sometimes, people give examples of incredible feats, which, allegedly, cannot be explained. They offer humanity's great achievements as proof that something mysterious and inexplicable must be happening in our minds. How did Mozart write his symphonies, and how can they move their listeners to tears? How did Goethe write Faust? These questions are rhetorical and meant as a challenge that cannot be met, the implication being that these things are so deep and complex that we cannot possibly explain them. That is an appeal to the supernatural.

Further, since we could in principle observe whatever it is Mozart and Goethe were doing when they wrote their great works as a process occurring in the universe, it is explicable. What they did was *create knowledge*; inexplicit knowledge, no doubt complex, perhaps knowledge they could not have put into words very well, but they were nonetheless creating knowledge in response to problems. Creating a piece of art that can move you to tears is not a spiritual endeavor. It is and always has been a software-engineering problem. Claiming that these great achievements cannot be "reduced" to "mere engineering" is a violation of computational universality. These things are mysterious only because we cannot explain them yet, but we should not conclude that they are inexplicable or spiritual. Creativity is behind all of them. One day we shall know how they work.

But we still have not exhausted the many facets of creativity denial. Surely, some say, something as straightforward as understanding an uncontroversial statement in conversation or watching a movie does not require creativity: you just need to listen or watch. This misconception is known as "instruction from without" (Popper 1996, 7-8), and it is induction again. It is impossible. What and how do you learn when you, say, watch a movie? How can you understand its plot? Does it take mere observation, or does it require active problem solving? The answer is the latter. It begins with the problem the film as a whole is meant to solve for you: perhaps you decided to watch it because you were bored, or because you were interested in it after hearing good things about it, or because your friends recommended it. As you watch, the film bombards you with a plethora of problems. What does this image mean? What is that character trying to say? If you do not solve these problems continually, you will not understand the movie.

Most of these problems may not be very hard, and you can solve them without great effort, even subconsciously. But sometimes it may not be clear what the movie displays on the screen, and you need to change your perspective a little bit in order to understand it. You are the active re-creator of the movie's meaning. Even seemingly uncontroversial depictions – say, a scene showing a tree – only seem uncontroversial because you created knowledge of what they represent beforehand, as a child. They are uncontroversial because they fit your background knowledge. The film may even present new problems you become interested in. That is what a good movie (or book or radio show) will do.

Sometimes knowledge *seems* manifest: imagine being a child and being in the kitchen with your brother. He puts his hand on the hot stove and cries out in pain. You see your brother's reaction and seem to understand *immediately* that the stove is something to be avoided. But that is not so. As a small child, you previously created all kinds of knowledge *from which* you can draw such a conclusion. You know that your brother is a human being; you know that *you* are a human being, too; you know that what hurts one human also hurts another, and you know that the expression on your brother's face and his scream are indicators that he is in pain. This entire situation unfolds lightning fast, so it appears as though there was knowledge manifest in the situation that you just had to observe. But some children may not yet know that what hurts one human may hurt another, and will curiously try placing their hand on the stove right after their brother did to see if it would hurt them, too. Not doing so is not a matter of age, but knowledge.

Seeing and identifying something can be virtually instantaneous: you look at an object – say, a guitar – and you immediately recognize its shape and know what it is. This immediacy can, again, make it appear as though knowledge is manifest, present, and just needs to be observed. But that is only in hindsight: at some point, you learned how to recognize a guitar. That took creative effort. You were not *born* with that knowledge. Now that you know, you can just invoke that knowledge, which does not take creative effort: running a guitar-recognition function requires no creativity, but writing that function in your mind in the first place does. And just because that function runs lightning-fast does not mean induction is true. The same is true in conversation: you are never really "told" anything, but actively create meaning. When somebody tells you, say, "Nice to meet you," you do not directly hear those words spoken. Sound waves travel through your ears into your brain, where conjectured functions convert them into sense data, which are then passed through a complex string of additional functions for you to even hear the words spoken. Then your mind invokes another string of functions in order to understand them.

In the "Oxford Companion to the Mind" (1987) in the section "Blindness, recovery from," the empiricist John Locke is said to have wondered in 1690: what would happen to a man who, having been blind his whole life, had his vision repaired? If a blind man had learned to distinguish a cube and a globe by touch, could he, once made to see, identify them just by looking at them? The "Oxford Companion to the Mind" goes on to give this touching account of a boy who went through this experience: "These remained interesting speculations, until in 1728 an unusually expert and thoughtful surgeon, William Cheselden, reported such a clinical case. Though generally distinguished as a surgeon his achievements were especially ophthalmic operations for cataract [...]. In a celebrated case Cheselden gave sight to a boy aged thirteen or fourteen who was born with highly opaque cataracts. Cheselden reported that:"

> When [the boy] first saw, he was so far from making any judgment of distances, that he thought all object whatever touched his eyes (as he expressed it) as what he felt did his skin, and thought no object so agreeable as those which are smooth and regular, though he could form no judgment of their shape, or guess what it was in any object that was pleasing to him: he knew not the shape of anything, nor any one thing from another, however different in shape or magnitude; but upon being told what things were, whose form he knew before from feeling, he would carefully observe, that he might know them again; and (as he said) at first learned to know, and again forgot a thousand things in a day. One particular only, though it might appear trifling, I will relate: Having often forgot which was the cat, and which the dog, he was ashamed to ask; but catching the cat, which he knew by feeling, he was observed to look at her steadfastly, and then, setting her down, said, So, puss, I shall know you another time.

The surgeon goes on to say that the boy was surprised that the things he had liked best when he was blind were not the most visually appealing; that those he loved were not the most beautiful.

It took him roughly two months to learn to recognize solid objects in pictures that he thought to contain only colored surfaces. Pictures gave him great trouble, as the report explains:

> [The boy expected] the pictures [to] feel like the things they represented, and was amazed when he found those parts, which by their light and shadow appeared now round and uneven, felt only flat like the rest, and asked which was the lying sense, feeling or seeing?

When shown a picture in his mother's locket, he recognized his father in it but could not fathom how a thing as large as a face could be displayed on so small a surface. The report includes an account of how many errors the boy had to correct to improve his shape-recognition algorithm. Things he saw initially appeared very large to him until he saw things even larger: while conceptually knowing that the room he was in was only part of the house, he could not imagine the whole house looking even bigger.

The boy initially expected little to come of the cataract surgery other than learning to read and write. He thought blindness gave him the advantage of being able to navigate places in the dark, and even after his surgery, he did not need a light to walk around his house at night. Nonetheless, his eyesight gave him great pleasure:

> He said, every new object was a new delight; and the pleasure was so great, that he wanted words to express it; but his gratitude to his operator he could not conceal, never seeing him for some time without tears of joy in his eyes, and other marks of affection. [...]

If knowledge came to us through the senses, the boy would have been able to understand and distinguish shapes visually immediately after his sight-restoring surgery. Things we take for granted – depth perception, for example – are complicated and learned concepts. The boy first had to *learn* how to use his new-found vision before he could learn anything else from it. If knowledge of how to use the senses were encoded genetically, he would not have had to do that. Therefore, knowledge of how to use the senses is neither encoded genetically nor can it possibly come from the senses, for that would lead to an infinite regress. So, where does it come from?

Once the boy got better at using his vision, he could learn from it by using it as an *error-correction* device. That is what senses are: error-correction devices. We use them *after* creatively conjecturing solutions to problems, such as when the boy picked up his pet to test whether it was a cat or a dog through touch. If he ever forgot, it would take more error correction, until he got good enough at it.

Something that seems as simple as learning to see represents a mountain of problems. Solving them is an impressive creative feat. Remarkably, babies and young children seem to do all of this effortlessly. They are confronted with floods of all kinds of sense data rushing in, yet joyfully manage to understand them, solving a myriad of problems simultaneously, integrating their senses into their understanding of the world so they can use them as skillfully and effortlessly as they do later in life. They learn to walk from virtually nothing, and they even learn to speak their first language while simultaneously learning how to use their vocal cords and mouths to make sounds.

In short, the surgeon's report is best explained as the boy having to conjecture theories at runtime with which to interpret, or rather, explain the newly received sense data. We might casually say that the story of the blind boy "shows" this. But that would be induction. Instead, what we mean when we say "the story shows," or generally, "the data show," is that the only good explanation for the story/data we have found so far is the one we have conjectured. We merely take a linguistic shortcut and say "the story shows," but it is only our conjectured interpretation of the story that can show anything.

The boy used his creativity to implement his visual shape-recognition algorithm. That was an impressive creative act. While it did not and cannot follow a predefined mechanism, it does follow a *logic* that I will now discuss. For, understanding this logic will allow us to build AGI because a newly created AGI will find itself in much the same boat as a baby learning how to make sense of the world around it for the first time.

The Logic of Problem Solving

The inventor Thomas Edison wrote: "None of my inventions came by accident. I see a worthwhile need to be met and I make trial after trial until it comes. What it boils down to is one per cent inspiration and ninety-nine per cent perspiration." We can divide all mental activities of people into two corresponding phases. The first is the inspiration/creative phase, in which we detect problems and then iteratively conjecture solutions and criticize them. Over time, through many iterations of error correction (with which the senses can but need not assist), we may come to a solution we tentatively consider good enough. The second is the perspiration phase, in which we apply the solutions we created. For example, once the boy has sufficiently corrected his errors in visually recognizing his cat, he will, from then on, have no effort applying this knowledge (i.e., running this function) to identify it again in the future. No further creativity is required – unless he finds more errors with this knowledge and decides to correct them.

We understand the inner workings of creativity to some degree. I will investigate them with the help of the example of a dead rat from Popper's essay "The Bucket and the Searchlight" (1983, appendix I), part of which I will now summarize.

Imagine yourself confronted with the following problem: there is a dead rat on the floor. You would like to know why. The rat being dead is the thing that needs explaining – it is the *explicandum* (Latin for "to be explained"). Nothing in the observation of the dead rat itself suggests a solution to this problem, so in order to explain what happened, you need to *guess* an explanation. For example, you may guess that the rat ate rat poison. However, from this sentence alone, you cannot validly deduce the *explicandum*. You need to combine the sentence with some *universal law* that you also guess, for example, as Popper suggests: "If a rat eats at least eight grains of rat poison, it will die within five minutes." Now you can combine this conjectured universal law with the initial condition, per Popper, "This rat ate at least eight grains of rat poison, more than five minutes ago." Together they constitute the explanation, and they allow us to deduce validly the explicandum that the rat has died. This combination is an example of a successful explanation.

Why is the statement "this rat has eaten rat poison" not sufficient on its own? Because, as Popper points out, you can imagine living in a world in which, when rats eat rat poison, they instead go on to be much stronger and healthier than before. In such a world, the universal law would be different. So an explanation that does not utilize at least one universal law is incomplete. (Instrumentalists claim that no such inclusion of a universal law is required.)

Thus, we get the following schema of deduction:

A universal law *U* + specific initial conditions *I* ———> *Explicandum E*

Since *U* and *I* together constitute the explanation, we can simplify the schema as follows:

Explanation T ———> *Explicandum E*

That *T* contains a universal law does not mean that every explanation is a law of physics or a law of nature (though *some* are). It means that *T*, when applied to the same problem situation, solves the problem *every time*.

As Popper continues to elaborate, a friend, skeptical of your explanation, might ask, "How do you know this rat ate poison?", thereby questioning the specific initial conditions you guessed. Alternatively, he might ask, "Even if it ate the poison, why should it have died of it?", thereby questioning your conjectured universal law. You must not reply, "How can you doubt it, seeing that it is dead?". Citing the *explicandum* as evidence for the explanation is circular, for the *explicandum* is what you are trying to explain in the first place. We instead require that the explanation make additional predictions that are independent of the *explicandum*. One such prediction, as Popper explains, is that if you were to inspect the contents of the rat's stomach, you would find rat poison. And the explanation may make many other independent predictions that we can use to test it.

This knowledge about explanations gives us the following schema (rotated for convenience so that it reads from top to bottom):

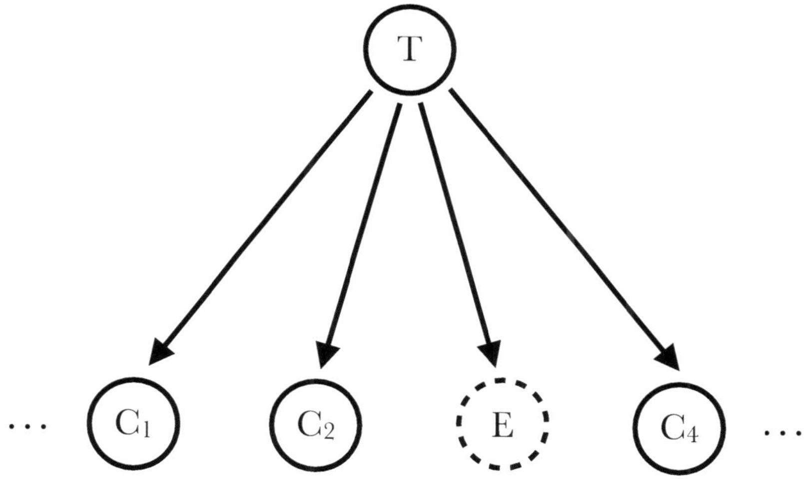

The explanation (or theory) *T* produces many claims *C*, of which claims one is the *explicandum E*, marked differently because it represents a problem. To deduce a claim, we invoke the theory with specific initial conditions. This is starting to look a lot like a function, is it not? To deduce something from a theory requires invoking that theory. This process is the same as invoking a function because the theory is a function. The specific initial conditions are inputs to the function, and the function's return values are its claims. We can now talk about explanations, theories, and functions interchangeably. *Using* a *theory* to *deduce claims* is precisely synonymous with *invoking* a *function* to *create return values*. It was when I first read Popper's explanation of the interplay between universal laws, specific initial conditions, and explicanda that it clicked for me and I conjectured that explanations are functions.

Invoking a function to create claims goes from top to bottom and is part of the perspiration phase. Given the specific initial conditions, you do not require creativity to do this. You may have noticed that I switched to calling it a "claim" rather than a "prediction" because *T* is not necessarily a *scientific* theory. Predictions are a small subset of all possible claims. *T* could be a theory about morals, in which case each claim says what to do next, or it could be a mathematical function, in which case each claim is either a number, a variable, or another mathematical function. Some theories return predictions in addition to other kinds of claims. *T* could be any kind of theory: this analysis of explanations applies universally.

The set of all applicable inputs to a function is called the function's *domain*, while the set of all possible return values is called the function's *range*. The specific initial conditions *I* are equivalent to the domain and the claims are equivalent to the range. Both the domain and range can be infinitely big. Implicit in (but not equivalent to) a function's domain and range are the problems it solves; that is, the function's reach. Infinitely large (and even finite but large) ranges cannot be explored exhaustively, which is why the implications of our knowledge can sometimes surprise us.

Following Popper (1983, 352), we can think of a specific set of inputs to a function as the *cause* of a particular return value or *effect*. However, we run the risk of slipping into instrumentalist terrain if we forget that a cause can only produce an effect in combination with the requisite theory. That theory (or law, explanation, or function, whatever we want to call it) is the necessary "logical link," as Popper calls it, between cause and effect, so we may instead decide to refer to a cause as the combination of a theory and a particular set of specific initial conditions. That is because the same cause can, through a different theory, lead to a different effect. Instrumentalists want to do the impossible: create return values without invoking a function. They do not realize this impossibility because they fail to appreciate the equivalence of functions and explanations.

To explain a problem is to create an explanation from which we can deduce it. We do this by guessing (Popper 1983, chapter 1). Once we have guessed an explanation, we can invoke it to produce additional claims by going from top to bottom in our schema. I guess this is what happens in our minds when we explore the consequences of our theories: we invoke them with many different inputs to see if they create problems.

Our schema also gives us a guideline for how to *test* a theory. For example, a scientific theory can be tested by deducing many predictions from the theory and then comparing them with observable, physical reality. If a prediction does not agree with reality, we have found a problem, and our explanation is refuted. The refutation goes from bottom to top: a single false prediction renders the entire theory false by *modus tollens* (Popper 2005, 18). The opposite is not true, however: finding that a prediction *does* seem to correspond to reality does *not* mean the theory is "verified." At most, we can claim to have "corroborated" the theory (Popper 1983, 17-21); and even this we should only say if we did not expect such a corroboration, that is, if the theory passed a severe test: a serious attempt to refute it. Predictions that are relevant to the test of a theory we call "risky predictions." (Popper 1983, 354) These are predictions that should be false if our *background knowledge* (Popper 1983, 71) is correct – that is, if other theories we currently deem uncontroversial (i.e., non-conflicting) are true. So to test a scientific theory, we invoke it with many different specific initial conditions to deduce several predictions, preferably risky ones, which we can then compare with reality.

But here we run into a problem: we do not have direct access to reality, so no direct comparison between a theory and reality can be made in order to detect problems. As the example of the boy who recovered from blindness shows, and as Popper said, even the most "direct" observations are "theory-laden," so things we "see right in front of us" we know only from theory. Likewise, if the theory we want to test is not scientific but instead of a different type, even direct access to physical reality could not help. Yet we do detect moral, mathematical, philosophical, and other problems all the time, even though they are all independent of the physical world. We even find problems that transcend a particular domain: a theory of what we actually did in physical reality may conflict with a moral theory of what we should have done, leading to regret. How do we do this?

To see how, we need to fix our schema, which is simplified and incomplete. Instead of trying to compare a theory's claims with reality, we need to compare them with claims deduced from another theory. When testing a *scientific* theory, the other theory is an interpretation of sense data. However, generally speaking, a problem is a conflict between *two or more* theories – scientific or otherwise – which is why our schema above is incomplete. We can only know that a particular claim is problematic if it conflicts with another theory. To return to our example, the dead rat is problematic only if you do not expect a dead rat. If a friend watches your house while you are gone and, upon your return, tells you "There was a rat, and I poisoned it, but I have not disposed of it yet," you will expect to find a dead rat in your house – it is now part of your horizon of expectations – and so finding it will not be problematic.

We thus extend our schema:

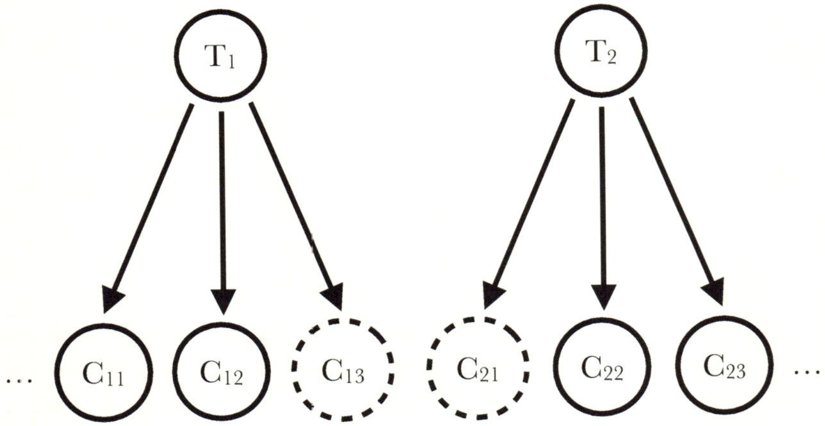

Now we have theories *T1* and *T2*. They produce the conflicting claims *C13* and *C21*, which, taken together, imply the *explicandum E*. If we study both theories carefully, we may find that they return many more than just one pair of conflicting claims. Sometimes an interplay between several more theories is needed to produce conflicting claims. So this is still a simplified schema.

To return to our example, *T1* is the conjectured theory of, say, what to expect when you get home. It predicts many things, partly based on your memory of what your house looked like when you left it. It predicts, for example, that your plants need watering because you have been gone for some time (*C11*; this is an example of an interplay with yet another theory that tells you that plants need watering after a while and why); that your mailbox is full (*C12*); that there are no foreign objects on your floor (*C13*), and many other claims. As you enter your home, you see a dead rat on the floor. Immediately, you identify the rat and recognize the problem. You do this so rapidly and intuitively it does not feel like there is much creative work happening. But not so fast: *T2* is an interpretation of sense data – itself conjectured in response to the problem of what you see in front of you at any moment – from which you deduce the presence of a dead rat (*C21*), the floor (*C22*), and some other, unrelated things in your visual field (*C23* et cetera).

Yet here we run into another problem. The claims "there are no foreign objects on the floor" and "there is a dead rat on the floor" do not directly conflict. Even to detect a direct conflict, one needs to have a repertoire of fairly sophisticated explanations; no-one is born with knowledge of logical rules. The problem *seems* manifest: as you read these claims, you immediately make the correct association and identify the problem. But again, we should not let the impressive speed and intuition with which we identify problems fool us into thinking that there is no creativity involved. And even if these two claims did *logically* contradict, how would we know? How would we detect that? And once detected, how would we know that it is something that needs fixing?

We need a third theory, also conjectured, which alerts us of the problem. It takes the claims *C13* and *C21* as specific initial conditions and returns a value signifying a problem. In other words, it explains why and how the two claims conflict. One possible such conjectured theory is "dead animals are foreign objects, so the floor does have a foreign object on it." Another is "the presence of dead animals is undesired," associated with the various requisite theories about preferences. So the detection of a problem, let alone its solution, is never goal-directed. It can only seem so in hindsight.

With the help of such a theory, we extend our schema once more:

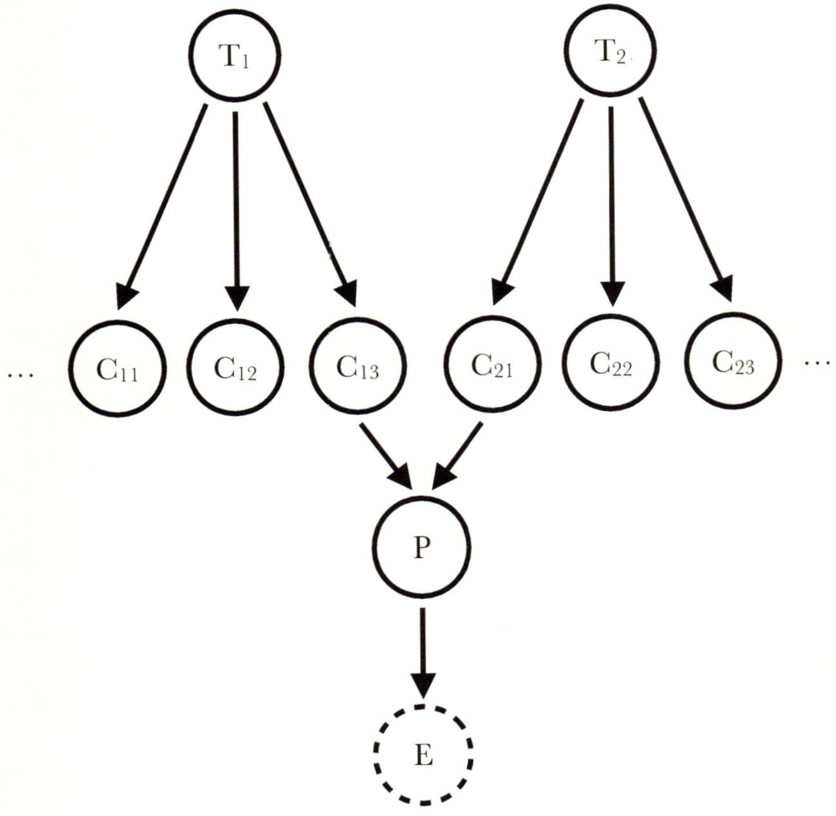

The introduction of a problem theory P allows us to deduce the existence of the *explicandum* E by invoking P with the allegedly conflicting claims. In other words, a problem theory is an explanation of why and how two (or more) theories conflict. Once we create P via conjecture, we try out different claims and then invoke P with them so that it tells us whether or not (and how and why) the claims conflict. When they do, we get an *explicandum* E.

The problem theory P may produce other claims (problematic or not) when invoked with other specific initial conditions, but that is not important here.

Can a *single* theory's claims conflict? Yes: that is a special kind of problem. It means there is an inconsistency within a theory. However, even that is only detectable if we have a theory/criterion of how to detect inconsistencies and that they are problematic. So a second theory is required even in this case.

The introduction of P and deduction of E together constitute an *understanding of the problem situation*. It is itself conjectural and subject to criticism. It has to be: sometimes we are mistaken about something being problematic. Likewise, the vast majority of problems we never discover, but if there were a way to find them mechanistically, we could (and would). Should a bad liar tell you, "I called the restaurant, and they told me they are closed forever," you may or may not notice a problem with that. Even with P present, we still need to iteratively try out different specific initial conditions to see if they lead to a problem. Indeed, we should expect most if not all of our knowledge to be incompatible, but not to have found problems yet because we have not made the requisite conjectures. That is the core tenet of fallibilism.

Now that we understand the problem situation, we can attempt to solve it. To find a solution is to conjecture an explanation that does not have the conflict; that is, it does not cause the same problem. But not any solution will do: "god put the rat there" does solve the problem, but it is not a *satisfactory* solution. In other words, we have criteria of what makes a good solution which we can use to criticize candidate solutions.

Several such criteria have been put forward throughout history. Authoritarianism generally and religion in particular require that the solution be consistent with authoritative texts. Induction and *Bayesianism* require that it be likely to be true; *foundationalism* that it be based on evidence; *logical positivism* that it be verifiable to be meaningful. Some misconceptions require it to be moderate (in the middle between two extremes, such as Aristotle's "mesotes," which states that we can find virtues between extremes – a variation on relativism). We separately refuted *those* criteria, also through conjecture and criticism, because we know of a better one: the solution needs to be a *good explanation*, that is, hard to vary (internally and externally) "while still accounting for what it purports to account for." (Deutsch 2012a, 31)

The search for good explanations implies other criteria. It implies the criterion of testability in scientific explanations, without which our scientific theories would be easy to vary while not matching interpretations of observations. It implies that we want our solution to preserve the two conflicting theories' explanatory contents except for the conflicting claims; for otherwise, our new theory would be easy to vary while violating our previous theories. It also implies that, should we decide to trade some explanatory content for another, we need an explanation for why we are doing so. Otherwise, we could pick any replacement for any explanatory content. That is what it means for a theory to live on in its successor as an approximation; it can also be phrased as the successor theory "simulating" the mistaken theory, such as general relativity simulating Newtonian physics. In other words, general relativity explains through the curvature of spacetime what looks like Newtonian physics. Likewise, Darwinism simulates Lamarckism (Popper 1983, 149). (And, on that note, neo-Darwinism simulates Darwinism.)

Another implication of the search for good explanations is that we will not accept a solution that introduces more or worse problems than the ones it solves. That would make the explanation easy to vary while still not solving those new problems, and so it would become a bad explanation. In other words, we have additional criteria about what types of problems we consider an improvement over previous problems, and a new explanation may be easy to vary while not meeting those criteria. Lastly, we want the new theory to fit our background knowledge, as well as avoid any infinite regresses, internal inconsistencies, et cetera. These criteria, taken together, make up our philosophy. They are searchlights that help us find errors.

The explanation tells us what led to the problem. It contains an identification of the mistaken theories. Sometimes, when two or more theories conflict, only one of them is false, but other times, some or even all of them are false. The identification of the mistaken theory or theories can be counterintuitive: when finding a conflict between a scientific theory and an interpretation of an observation, it is not always the scientific theory that is false. Sometimes the interpretation of an observation is false, as in the case of optical illusions.

If problems, solutions, and criticisms are all conjectured – is it not all subjective? How can this process possibly come closer to the truth? While these things are indeed all conjectural, it is whenever we happen to be correct about them that we make real progress. How could it be otherwise? There is no possible way to prove with certainty that a problem exists, or that a solution works, or that a particular criticism should be applied. They must be conjectural because sometimes we think we have found a problem and then, after further investigation, find that it is not a problem after all.

Likewise, we can be mistaken in our criticism of a theory, or our solution to a problem. But when a solution *does* happen to work, it is because at least part of it corresponds to reality. Our airplanes fly because our theories of aerodynamics contain some truth. If tomorrow we find out that our theories are mistaken, planes will not suddenly crash. And when we happen to identify a conflict between theories correctly, we really do see an opportunity for progress, because, as I have said, no such conflicts exist in reality. We get those parts right that we happen to get right, and through an effort, our searchlights may help us get more things right.

We are presented with another problem: there are infinitely many theories that solve the problem, meet all of our criteria, and are nonetheless all false. But this is again no reason for despair, nor for a descent into subjectivity: we can only guess so many theories, so we do not need to deal with infinite amounts of them. And if the best one of them, despite surviving rigorous criticism, is later found to be false, that is an opportunity for further growth. It means our error-correcting mechanisms are not rigorous enough, and we need to come up with better ways to criticize: we need to conjecture additional criteria.

On the note of subjectivity: Building on Deutsch's findings in aesthetics (2012a, chapter 14), I submit that there is no such thing as a purely subjective theory in the sense that there exists a matter to which there is no objective truth. I guess that appeals to subjectivity are the result of having bad explanations of the applicable domain. Even in scenarios that seem dictated only by taste, such as one person liking cheese and another disliking it, that is a problem, and there is some truth to be found. Perhaps a better kind of cheese could be made that both like, whereby we create knowledge about culinary tastes and learn several biological truths along the way, too. Just because we do not yet have good explanations of how taste works, that does not mean it is subjective. Problems indicate objectivity. Wherever there is something to be known, we can make objective progress.

With the help of the latest extension to our schema, we can now iteratively conjecture solutions and subject them to the criticism of our criteria until, hopefully, at least one conjecture emerges as a tentative solution. Depending on the difficulty of the problem and our background knowledge, we may no doubt do much of this quickly and subconsciously. But we still do it – there is no process of inferring a solution, problem, or criticism. That would be induction. (Inductivists claim there is an impossible line in our schema that takes them from a problem below back up to an explanation somehow.)

Following the logic I have outlined, we can arrive at a solution such as: "My neighbor has rats and has been trying to poison them. The poison kills rats that eat more than eight grains of it within five minutes. One of his rats must have eaten enough of the poison, crawled out of his house and entered mine within five minutes to look for a place to hide, where it died." This solution meets all of our criteria. It is no *proof* that this is what happened – we can only ever guess.

While invoking a solution goes top-down, finding a solution is an attempt to place a new theory at the top. Since nothing in the problem situation below itself suggests a solution, we cannot go bottom-up, but need to instead try out different solutions by placing them above and seeing if they work. This phase is the inspiration phase: it is the creative part. It cannot be automated in the sense that it is always *guaranteed* to find a solution, but in principle, it *can* find a solution to any soluble problem. This relationship is inverted in some contexts of statistics, where a given model is said to "explain" a data point because that data point can be deduced from it (Deutsch 2012a, 15). But this has nothing to do with explaining. In reality, explaining is the process of placing a conjectured theory *atop* an *explicandum*. Explaining must go bottom-up.

When a solution is found without any outstanding criticisms, we place the new theory – *T3* – at the top of the schema and thereby replace the previous theories.

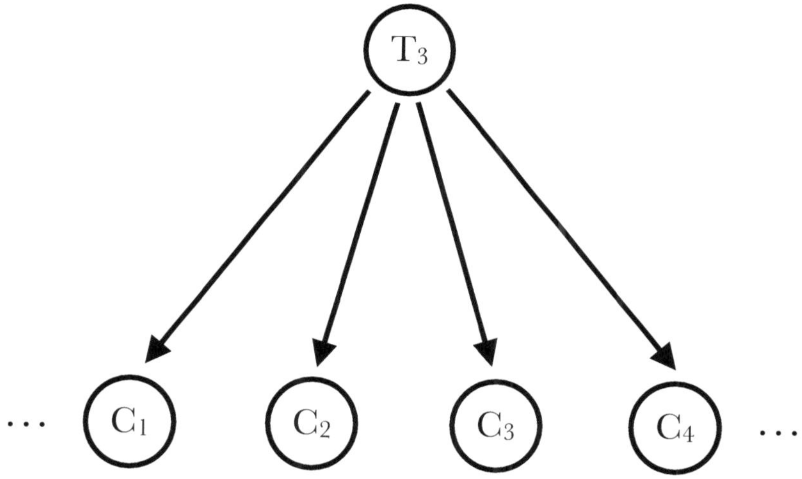

At this point, we tentatively consider *T3* unproblematic; but in time, we may find a new, better problem with this new theory. We see now that solving a problem as simple as "why is there a dead rat on the floor?" requires a complex interplay of all kinds of conjectured theories and criticisms. It requires vast amounts of background knowledge about rats, floors, rat poison, how these things may conflict, and many other things. The entire process follows this logic:

1. Find a problem
    1.1. This step requires that you find a conflict between theories; that is, a conflict according to some previously conjectured criterion. Repeatedly invoke two or more theories by creatively trying out different

specific initial conditions and feed the returned claims into the criterion

1.2. If the claims are conflicting, that is, problematic, the criterion will alert you

1.2.1. If it does not alert you, try different specific initial conditions, or try finding a conflict between other theories

1.2.2. If it does alert you, continue with number 2

2. Conjecture several different explanations and criticize them based on previously conjectured and currently uncontroversial criteria[13]

2.1. Repeatedly alter, combine, rearrange, and add to existing ideas to create a pool of tentative solutions

2.2. Invoke the tentative solutions by creatively trying out different specific initial conditions and feed the returned claims – ideally risky claims – into your criteria

2.3. If your criteria alert you of problems, go back to 2.1 until you have found a solution that matches your criteria.

3. To find new problems with your new, better theories, go back to number 1

In short, we can simplify our schemata to Popper's schema of problem solving (here slightly modified, compare 1983, 243):

P1 ——> TT ——> EE ——> P2

It reads: in order to solve the first problem *P1*, we conjecture many different candidate theories *TT*, then eliminate many errors *EE* to hopefully end up with a remaining solution, with which we may then find a second problem *P2*. This schema is still simplified: at various points in the process, we may detect additional, seemingly unrelated problems, toward which we can branch off in order to solve them. Several steps may require new branches of interim explanations, such as why something is problematic, why we should eliminate a solution, and why we do not need to preserve all of the original explanation's explanatory content. So, in reality, the whole thing looks more like a tree, but as a simplification, this will do.

In chapter 3, I discussed how evolution is a theory about the creation of knowledge generally, not just biological knowledge. And indeed, the process of conjecture and error elimination in the human mind is evolution: conjecture is the variation of an idea. It has its biological counterpart in the mutation of a gene, while criticism and testing have their counterpart in natural selection (Popper 1983, chapter 7), with the unit of selection being a single idea. The mind at no point passively receives instructions, but instead actively tries out different theories and weeds out those that are maladaptive while keeping the adaptive ones. The schemata above incorporate these parallels to biological evolution.

In biological evolution, variations happen *without* regard for the problem situation. The same is true of problem solving in our minds. Sometimes we find solutions quickly, but we only ever know that something is a solution *after* we created it. Its creation must likewise happen without regard for the problem situation. This fact is true even in situations in which we intentionally solve a problem. We do not think of a solution *in order to* solve a problem for the same reason giraffes did not evolve longer necks *in order to* reach higher leaves (or, as has been suggested, compete better with other males).

However, upon closer inspection of the evolutionary theory of human knowledge, we run into a problem. Evolution is a theory about *replicators* (Dawkins 2016, chapter 2): it explains the creation of knowledge as a result of replicators competing with each other in order to spread through a *population* of replicators. But so far, there is no notion of a replicator in the context of human knowledge within single minds. We know that *memes* replicate *across* minds (Dawkins 2016, chapter 11) – they spread by being copied from one mind to another – but if knowledge is to be created *within a single mind*, there has to be a replicator within that single mind as well. In other words, it is not enough to use evolution as an *analogy* of how human knowledge is created. Evolution is our only working theory of how knowledge of any kind emerges. Therefore, we need to explain the creation of human knowledge *literally* (Popper 1983, 261) in terms of evolution. Doing so requires a notion of replication within a mind that meme theory does not provide.

Secondly, we can detect several problems with the top-down schemata above: we presumably have a criterion telling us how long we are willing to spend alternating between conjectures and criticism. Where does it come from? And since every criterion of criticism must itself be the result of alternating guesses and criticism, we have an infinite regress: how did we criticize our criteria of criticism without having any criteria of criticism in the first place? There could not have been a "first" criterion, because we know this first criterion itself is knowledge that we must have created through conjecture and criticism. There could have been no other criteria to criticize *that* criterion.

We find another infinite regress in the fact that, in order to criticize a theory, we need to *creatively* try out different specific initial conditions to detect a problem with it, as well as creatively come up with a problem theory alleging the problem in the first place. So in both cases, creativity seems to depend on creativity existing beforehand. And we have not yet explained one of the basic building blocks of creativity: conjectures. Where do they come from? It is not good enough to explain conjectures in terms of varying an idea: that is just another way of phrasing the same phenomenon. Instead, I mean, how do variations of ideas occur, in detail?

All three problems – the missing notion of a replicator within minds to explain the creation of knowledge, the occurrence of multiple infinite regresses within the above explanation of the creative algorithm, as well as the mechanisms behind conjecture – have the same solution. That is because our schemata above are higher-level explanations of phenomena that emerge from an underlying process that does something else. But before I explain that underlying process, I must first make a brief detour through the territories of free will, consciousness, and animal "learning."

Free Will and Consciousness

I feel, almost grudgingly, that I should address the problems of *free will* and *consciousness*. They are some of the more prominent victims of bad philosophy. From time immemorial, philosophers have offered bad explanations for them, partly because these problems are hard, partly because philosophers have not had the requisite philosophy to correct their errors in their thinking about them. My treatment of them will not be new to Popperian readers. However, I think I should mention them because we can address the problem of free will relatively easily given the theories I have so far presented. And the problem of consciousness, while harder, relates meaningfully to the solutions to previously raised problems as well as the question of whether animals are conscious or not, which is part of the next section. More importantly, if this section's central thesis is correct, two things follow: first, neither free will nor consciousness is something we need to explain in order to build AGI; second, AGI will have both.

My central thesis in this section is simply stated as: both free will and consciousness are *epi-creative*. (Deutsch (2012a, 157) has made a similar conjecture regarding consciousness.) By "epi-creative" I mean they arise from the creative algorithm (and only from the creative algorithm) as a sort of byproduct.

I will first focus on the problem of free will. Free will is the idea that we are the authors and enactors of our choices. The problem of free will consists of two ideas that have traditionally opposed free will: *determinism* as well as *indeterminism*. Determinism says that all events are entirely determined by prior causes: that one thing follows another like clockwork. The laws of physics determine the history of the universe, leaving no room for free will. Indeterminism is the opposite view: it says that events are not caused by prior events, or at least do not follow any discernible causes. Instead, indeterminism says, events are the result of chance, particularly on the subatomic level, which chance, according to some interpretations, allegedly percolates to higher levels of emergence.

Whether determinism and indeterminism are true or false need not concern us here. The dichotomy of determinism vs. indeterminism has damaged inquiries into the concept of free will almost beyond recognition, for, once you are stuck within the fangs of this dichotomy, thinking you must choose between the two, or get them to agree, you will not realize there is another way, and end up creating only bad explanations as a result. We should replace the focus on the dichotomy with a search for good explanations.

Recall the criterion of reality (Deutsch 2012a, 23): "a particular thing is real if and only if it figures in our best explanation of something." If we want to explain how and why, say, Joe went to the ice cream parlor, then both "the laws of physics made him do it" and "chance made him do it" are bad explanations. They are bad explanations because, even if true, they are easy to vary externally: they apply to absolutely everything. A good – and common-sense – explanation is, "Joe went because he felt like eating ice cream."

Opponents of free will would no doubt reply, "We do not know how he decided *that*, though." Reductionists, for example, are particularly fierce opponents of free will. *Reductionism* is the mistaken prejudice that we should always explain phenomena in terms of lower-level components (Deutsch 2012a, 123), and that causality can only go from the small to the large, not the other way round.

Thus, reductionists might claim that Joe's actions should be explained solely in terms of electrical crackles in his brain: if the crackles happen to crackle a certain way, Joe will eat ice cream. But "his neurons made him do it" is just another bad explanation. Good explanations operate on the appropriate level of emergence (which, by the way, is why neuroscience cannot explain emergent phenomena such as consciousness, free will, or, most relevant for our purposes, intelligence; see chapter 7). Thinking that causality can only go bottom-up, from neurons to the mind, say, and not top-down, is a reductionist mistake. (Sometimes, while trying to explain something well, you may find yourself stuck on a single level of emergence. This situation can be due to reductionism. But whatever the reason, as a rule of thumb, if you cannot go left or right, try up or down, that is, change the level of emergence on which you try to explain something.)

In any case, we *do* know how Joe decided, because we can make good guesses about it. If we know Joe well, we know that he has loved ice cream ever since he was a child; that is, we explain his choice through common sense in terms of *preferences*: knowledge Joe created early on in life about what foods he likes. People follow software instructions they themselves created in their minds. That is what human knowledge is. Preferences are one kind of software. People creatively conjecture new solutions to problems, including choices they want to make, as well as criteria for when to make them and how. So we are very much the authors and enactors of our choices. We make our neurons do things.

There is an applicable deterministic criticism of free will on the appropriate level of emergence of software. If a person is an instance of the creative program, and programs are executed by computers, including the brain, blindly, step by step, without any room for error or deviation from the instructions, how can there be free will? To this I answer: it is correct that people are programs (Deutsch 2012a, 414), and that computers, such as brains, execute programs stubbornly. But this does not violate the concept of being the author and enactor of one's choices, for the creative program confers free will precisely *because* it is executed faithfully. And the errors people make are a result of conflicting ideas produced by their instructions, not deviations from the instructions (more on the difference between these two kinds of errors, and why following instructions does not preclude free will specifically, or creativity generally, in chapter 7).

A second applicable criticism of free will rests on our previous analysis of creativity: we detect problems and solutions in a roundabout way through an evolutionary process. Evolution is not goal-directed or purposeful; it is selectionist, not instructionist. But people with free will have purpose, meaning, goals, and intentions, so there is a conflict. To this, I answer: we must consider purpose and intentionality on two different levels of emergence. Within evolution in a mind – underneath the hood, so to speak

– it is true that there is no purpose. Blind variation and criticism create new knowledge, and, somehow, lead to an emergent "I," the self. But on the level of that "I," purpose, intentions, and goals do very much exist: they are a kind of knowledge that this "I" creates through just that evolutionary process. They are explanations of what to do and what to strive for that emerge from the underlying evolution.

In conclusion, if by "free will," we mean "a computer not following instructions" or "mechanistically finding problems or solutions" or "not being limited by the laws of physics," that cannot exist. If instead, we mean "the individual's future is open" and "we constantly rewrite our instructions by explaining the world, forming preferences, and making choices, independently of our material substrate, and these instructions are neither given nor set in stone," then free will does very much exist. We have free will because we *are* the authors and enactors of our choices.

With that, I consider the problem of free will solved.

On to consciousness. It is a profoundly mysterious phenomenon, not least because we cannot describe or predict qualia yet (Deutsch 2012a, 153). I do not pretend to have an explanation of consciousness. But even without one, I have criteria that allow me to criticize fashionable yet nonsensical explanations of consciousness, and so I must criticize them.

One thing you might hear thrown around is that consciousness is when the mind's model of the world includes itself, and various versions thereof, all centered around the theme of recursion. Recursion is when a function invokes itself and keeps doing so until it reaches some condition to terminate. I do not know why or how recursion got its status as a go-to woo-woo explanation for difficult concepts. Perhaps it sounds scary enough to deter people from asking further questions while making its originator sound smart. But in any case, people should never be impressed by bad explanations. It is trivial to build programs that employ recursion. One can build algorithms like factorials or Fibonacci sequences in terms of it. Indeed, all algorithms that use loops can be transformed into ones using recursion.[14] But it is doubtful that an algorithm that calculates factorials should be conscious.

Would a program that does something as mundane as, say, *looping* over a list of itineraries in order to print them out for a vacation, suddenly become conscious if it was changed to *recur* over that same list instead? One can even implement something as simple as *addition* in terms of recursion, so recursion by itself does not seem *sufficient* for consciousness, nor has anyone explained well why it should be *necessary*. In fact, we implemented addition in a recursive fashion in in chapter 3, and that implementation is *only* an explanation of addition, not of consciousness.

People often consider the ability to recognize oneself evidence of consciousness. Like Deutsch (2012, 154), I disagree with that conclusion. Recognizing oneself is just a special case of recognizing faces and other objects generally. We now have technology that is sophisticated enough to recognize all of them. Websites can automatically tag people in photos. Are these websites conscious? Again, I doubt it. Indeed, I guess that babies are conscious long before they learn to recognize themselves in the mirror.

Lastly, attempts to explain consciousness are sometimes made in terms of modeling one's position relative to one's surroundings (this is sometimes offered as an explanation of creativity generally). And once again, we find that people have already built algorithms querying the global positioning system (GPS), as well as algorithms that can navigate the interiors of homes. Drones can fly from point `a` to point `b` using navigation software, and automated vacuum cleaners can create maps of the interiors of homes to vacuum them (Temple, 2019b) and, when done, find their way back to their power station without human assistance. It is again doubtful that any of these programs are conscious.

Some do not attempt to *explain* consciousness, but instead, they *deny* that it exists. The denial of consciousness started in the 20th century and has its roots in both behaviorist psychology and reductionism (Strawson 2018). If we attempt to explain human minds merely as the cumulative result of conditioning, without need for thoughts or feelings, not only do we not *require* consciousness; it can *get in the way* of our behaviorist "explanations," for, now there is something significant *between* the inputs and outputs of the human mind that we need to explain. Behaviorism rests on the assumption that there is nothing in between to explain. Likewise, if we try to explain human minds merely in terms of electrical crackles in the brain, we only need hardware. Notions of software are in our way in this case, too, including consciousness.

While reductionism to hardware was a good defense against bad explanations involving "souls," "essences," and the like, it denies the existence of software, and of anything immaterial, for that matter, and so would never allow for a good explanation of consciousness (or creativity generally). *Dualism*, on the other hand – the true theory that both material and abstract things are real and interact – has long been known in software engineering as the distinction between hardware and software. (Although "pluralism" is a more fitting term since software lives on multiple levels of emergence.) That dualism is not controversial among software engineers is another example of how software engineering can help solve philosophical problems. Like all other software, consciousness really does exist.

There is a misconception that should receive an honorable mention – not because it is a halfway decent explanation, but because it is a mistake reminiscent of another one we have already encountered. It is the doctrine of *panpsychism*. It states that *every* object is conscious, at least to a degree, including plants and stones. It correctly rejects a reductionist hardware focus and offers a third alternative in addition to explanations of consciousness based on hardware and those based on software: it does not require such a distinction.

But in reality, panpsychism is not an explanation of consciousness for the same reason Lamarckism is not an explanation of the origin of knowledge. Recall that Lamarckism claims that new knowledge has somehow already been present in an organism, and, therefore, does not need to explain its origin anymore. Likewise, panpsychism claims that consciousness is somehow already present *everywhere*, and, therefore, does not need to explain its origin anymore, either. Not only does panpsychism not *explain* consciousness or its origin: it *denies* that it exists, just like Lamarckism denies the creation of knowledge. If *everything* is conscious, *nothing* is. I consider the fact that both

Lamarckism and panpsychism make the same underlying mistake evidence that consciousness is related to the origin of knowledge: error correction.

Which, following Popper (1983, 344), brings me to my main conjecture about consciousness. So far, I have relied on intuitive notions of consciousness to refute bad explanations of it. There is a better yardstick. We can make rough guesses about what gives rise to consciousness and what leads to its absence. I guess that consciousness is related to the search for and correcting of errors: you are most aware of whatever problem you are working on. For example, when you first learned to ride a bike, that was a conscious effort. You had to correct thousands of minute errors in your balance, your steering, and your pedaling simultaneously. As you got better at these things, that is, the more errors you corrected, the less aware you became of what it was you were doing. You began to steer and pedal automatically, but perhaps still had to correct errors in your balance. Once you straightened out most or all of those errors, you began to ride the bike effortlessly, as if on auto-pilot. I assume this is how you ride your bike today: it does not take much awareness to ride the bike itself. Instead, you focus your attention on, that is, look for errors in your surroundings to avoid an accident. When an error does occur – for example, you roll over a rock too big and too close to avoid – you become *acutely* aware of the situation.

If you play an instrument, there may have been times when you felt wholly immersed while playing, and realized after finishing the song that you had not noticed time going by, nor had you seemed to do any of it consciously. At least if it is a sufficiently well-practiced song: the more errors you have corrected, the less aware you become of playing the song.

When you reach the top of a flight of stairs, thinking there is one more step, you also become *acutely* aware of the situation. You had an expectation that was disappointed (Popper 1983, 344). Only when it is disappointed, do you become aware of your expectation and are given an opportunity to correct it. Learning consists of the elimination of disappointed expectations. Whenever you focus your attention on something, it means you look for errors with it. When you find an error, this, depending on the situation, sparks curiosity, or annoyance, or other related qualia based on context. Physical pain is an indication that there are errors in your body that need fixing. If you then succeed in solving that error, you experience joy or relief. The bigger the error was, the more aware you are of finding its solution, too.

While we are aware of *finding* a solution to a problem, we are not aware of *how* we found it, that is, the creative process that led to the solution. Even when intensely mulling over a problem, we are aware only of many intermediate solutions, but not the processes in between. We conjecture subconsciously all the time. That explains how we can have sudden epiphanies in the shower (known as "shower thoughts"), or wake up in the middle of the night with a solution to a problem. Sleeping may be a fairly uncreative state – one is unconscious because one is not creative, not the other way round – but should you happen to find or solve a problem while asleep, it can make you aware that you are dreaming, or even wake you up. You always direct your attention toward a problem you want to solve. You are aware of the problem, and, should you succeed in solving it, its solution.

I do not claim that *any* system that implements error correction is conscious. As Ella Hoeppner pointed out to me (personal communication, 2019), there exist error-correction mechanisms that we can explain without invoking consciousness. For example, gene replication is accompanied by an error-correction mechanism to avoid mutations. Likewise, computer programs can correct errors in faulty transmissions of messages. Neither is conscious because they are different from creative error correction in three ways. First, they cannot universally detect errors, but only those which they are programmed to detect. Second, they do not detect errors in the sense of conflicts between theories. Third, they do not detect errors in the creative sense: we can think of them more as "quality assurance algorithms." They change a given state to another given some criterion. For example, DNA mismatch repair is achieved by enzymes iterating over the base pairs in a newly replicated DNA strand to make sure each base is paired correctly: guanine must only be paired with cytosine, and adenine must only be paired with thymine. Since this repair happens according to a pre-defined mechanism, I do not consider it error detection and correction in the *creative* sense; it is more about finding a breaking pattern. So error correction in the conscious sense seems to involve the ability to find and correct *unexpected* errors creatively.

In any case, none of this *explains* what consciousness and qualia *are* or how they work, but it explains what gives rise to them: error detection and correction. Their absence is the reason the examples I gave of nonsense explanations are false; they do not involve error detection and correction. Now we have a yardstick for evaluating alleged explanations of consciousness: they can only be true if they explain consciousness in terms of error correction and detection. And, since it is error detection and correction that seem to give rise to consciousness, and since they are the key ingredients of creativity, I conclude that consciousness, like free will, is epi-creative. I further conclude – tentatively, as always – that consciousness is at least a property of software, or perhaps itself a piece of software.

Animal "Learning"

Studying how animals "learn" is interesting because it bears a resemblance to how narrow AI works (see chapter 7). Since, as I have argued, consciousness and free will are epi-creative, determining whether or not animals are creative also determines whether they are conscious and have free will. Lastly, an understanding of the inner lives of animals will shine a light on how creativity may have evolved.

Animals are a part of our lives. They are a part of the environment and our food supply. Some we eat, some we love. Some are our companions. This tight integration of animal lives into human lives brings us closer to animals, and the sophisticated behavior of some animals seems to be evidence for their creative and emotional capacities. To doubt these capacities has become a stigma and, so the sentiment implies, one must be a soulless monster to believe animals are not at least somewhat intelligent, let alone have the capacity to suffer. But what if it follows from our best explanations of intelligence and suffering that they have neither?

There have been philosophers who have described animals as non-sentient automata. Today, they are ridiculed and vilified. Most would think rightly so. Our dogs get excited when we come home; our cats purr as they sit on our laps. Bees display some of the most intricate mechanisms of communication in the animal kingdom. Elephants seem to mourn their dead.

Intelligence and consciousness are generally attributed to animal species in various degrees, depending on how distant an evolutionary relative of humans they are. Mammals are said to be nearly as intelligent and conscious as humans; fish less so; insects only a little bit; and plants and fungi not at all. In order to understand how consciousness arises gradually in different species, people have proposed various explanations in terms of the organism's physiology. Even Popper said we should grant that animals that have a central nervous system are conscious (a reductionist, hardware-biased mistake) and that consciousness exists in degrees (Popper 2018a, 63). While at any given moment, we are indeed more aware of some things than others, one has to be conscious first to be aware of anything at all. And explaining emergent phenomena such as intelligence and consciousness in terms of the underlying hardware is a reductionist mistake that violates computational universality.

Whether an animal (or a machine, or any other object) is intelligent depends on the *origin* of its knowledge. Since evolution is the only process capable of creating knowledge that we know of, we are looking for an evolutionary process. If the evolution of said knowledge occurs in the animal, during the animal's lifetime, then it is intelligent, creative. If it does not – for example, if all the knowledge in the animal is created by *biological* evolution *before* the animal's lifetime and inherited in genes – then the animal is *not* intelligent. The phenotype of an animal's genes includes not only the animal's hardware – its body – but also its behaviors. We readily grant that a dog's coat is genetically encoded, but it is a bit more challenging to see that its ability to, say, play fetch is also genetically encoded (see also Dawkins 2016).

Following Deutsch, I believe the simple fact of the matter is that all animal behavior can be explained well solely in terms of pre-existing, genetically given algorithms (personal communication with author, 2020; see also Temple, 2003a, 2019b, 2019c). They include tool use, self-recognition, and emotional responses (without the associated qualia – see Deutsch 2012a, 154). To infer from animal behavior that *looks* intelligent or conscious that animals *are* intelligent or conscious rests on the behaviorist mistake of coming to a conclusion about the animal's inner state based only on its expressed behaviors rather than an *explanation* of how animals work "on the inside." (We will re-encounter this mistake in a similar form in the context of the Turing test in chapter 7.) And we know that animals are not intelligent because it *follows* from the fact that they miss a genetic mutation that we humans inherit from a single one of our ancestors (which mutation I will explain in the next section).

In everyday life, we make a simplified and convenient assumption that animals create knowledge at runtime (i.e., their lifetime) in order to solve problems. That is what intelligence is. In reality, the gene pool is filled with genes that influence animal behavior in such a way that it *looks as though* animals create knowledge themselves.

We can explain changes in an animal's behavior during its lifetime in terms of positive and negative reinforcement. For example, a dog will generally repeat behaviors for which it is praised and reduce behaviors for which it is punished or scolded. The range of behaviors (the *policy*) it can exhibit is encoded genetically and cannot be deviated from, and therefore does not allow for creativity. To be rewarded, it can choose from those behaviors and combinations thereof. When successful, the solution has been "learned" or "remembered," or rather, "imprinted a little more deeply." Afterward, should the dog encounter a similar problem, it tries all previous behaviors, or at least hints of them (including the unsuccessful ones) in the same order and rapid succession, until it reaches the praised behavior. Over time, its behavior will be dominated by rewarded actions, while unsuccessful behaviors are reduced to mere indications until they almost disappear (Popper 2002a, 17).

One of the more well-known examples of an allegedly intelligent animal is Koko, a female gorilla. She was born at the San Francisco Zoo in 1971 and spent most of her life at a gorilla preserve in Woodside, California, where she died in 2018. She checked all the boxes: biologically speaking, she was a close relative of humans; her physiology is similar to that of humans; she could use her face and arms to express emotional responses, and, most impressive of all, she "learned" a modified version of American sign language. When the actor Robin Williams met with her, he called it a "mind-altering experience." During a visit from television personality Mr. Rogers, she signaled to him that she loved his visit. You might not believe your eyes, until right after Koko said it, Francine "Penny" Patterson, figurative mother and teacher of Koko, *praised her*. And that was the mistake. Videos are available online of Patterson communicating with Koko and praising her whenever Koko gives an answer that makes sense and repeating or rephrasing a question or command whenever it does not.

In other words, *Patterson* corrected Koko's errors, not Koko herself. I do not doubt that Patterson felt an emotional connection to her. Koko may even have reciprocated in the form of emotional responses. But many will be disappointed to learn that they did not "come from the heart," and could not have. All exchanges with Koko, including emotional responses, were a result of years of conditioning and reinforcement. There was no intelligence, no creativity at play. Her behaviors changed, but no new knowledge was created within Koko.

These mistakes could have been avoided. In early 20th century Germany, at a time when people started to become interested in animal intelligence, the math teacher and amateur horse trainer Wilhelm von Osten owned a horse named Clever Hans.[15] Von Osten claimed to have taught Hans how to do basic arithmetic and understand basic German sentences. He would ask the horse, say, "What is 5 + 5?", and Hans would tap his hoof ten times. Not long after von Osten began to travel the country to show off his horse, drawing large crowds at every performance, scientists became interested in Clever Hans. They first performed experiments that are best explained as evidence that Hans usually gave the right answer when he could see whoever asked him a question. The inquirer did not have to be von Osten, meaning he was not communicating secretly trained signs to Hans.

Next, Hans was provided with blinkers so he could only look straight ahead. If he could see the inquirer, he got the answer right almost every time. But whenever the inquirer was standing to the side, invisible to Hans, then Hans almost always got it wrong, even though he could hear the question. The scientists further noticed that Hans started to tap more slowly as he approached the correct answer. He was watching the inquirer and looking for a signal to stop. Von Osten must not have realized he was giving Hans signals just by looking at him a certain way. Eventually, the scientists figured out that they could get Hans to give any arbitrary response, right or wrong, by *raising their eyebrows!* That was the signal Hans was waiting to detect. At the beginning of his training, von Osten presumably got excited every time Hans got closer to the correct answer, praising him and expressing the excitement in his face by opening his eyes wide, which included raising his eyebrows. Hans then associated having done the right thing with raised eyebrows.

Therefore, the whole process can be reduced to reinforcement again, which, as I have said, works according to genetically predefined mechanisms that do not require creativity. Horses are domesticated animals, so it would not be surprising if genetic mutations had been favored in the past that made horses more responsive to people's facial expressions. So Hans' ability to recognize and interpret raised eyebrows may well have been genetically given. It was a neat party trick, but Hans did not know what he was doing or why.

After the Clever-Hans fiasco, many experimenters in the studies of animal cognition learned the lesson and strictly avoided any face-to-face contact with their animal subjects. But this lesson is still not widely applied, see Koko. Indeed, hardly anyone seems to notice that Koko could not imitate, let alone understand, spoken language, despite the complex inborn calls that apes can make. I suspect this is the reason Patterson communicated with Koko through sign language rather than mimicking ape calls.

Communication of any sort, no matter how sophisticated, is no evidence of intelligence. Honey bees display vastly more sophisticated ways of communicating in their minute dance than Koko ever did. Through an intricate dance pattern, a successful forager bee communicates to other members of the hive the location of pollen, water, and new nesting sites. The exact details of the dance are too subtle and numerous to mention here, but suffice it to say that the duration of a part of the dance indicates the distance to the destination, and the bee's movement at an angle relative to the sun's rays (!) indicate the direction towards the destination.[16] Knowledge is no doubt present in all of these cases – otherwise, these animals would not be able to do any of the things they do – but it is given *genetically*. It is either its sophistication or its leading to changes in behavior that are mistaken for intelligence. The extent to which behavior can change should not impress us: some reinforcement algorithms have reach.

How come animal behavior is sometimes similar to that of humans? For example, many animals and humans share similar responses to pain: generally speaking, both try to escape the pain's source. Since humans are intelligent, is this shared behavior not evidence of intelligence in animals? No. Again, it is the *origin* of knowledge that determines whether an entity is intelligent, not any particular knowledge. In this case, the behaviors are similar because the animal's and human's goals are roughly identical (Hoeppner, personal communication with author, 2019). In the animal, knowledge of how to escape pain was created by biological evolution: those genes that happened to instruct the animal to avoid sources of pain were able to spread through the population. On the other hand, a human in pain creates this knowledge through the evolution occurring in his mind. Since both biological evolution and the evolution in a human's mind are evolutionary algorithms, Hoeppner argues it can make sense for them to converge to similar solutions for similar problems.

What *would* constitute evidence of intelligence in animals (and organisms and programs of any kind)? Behavior which, in front of the background of our best explanations, shows that it is the result of knowledge that is not encoded in the animal's genes, but was instead created by the animal itself during its lifetime, as well as knowledge for which there would have been no selection pressure in the evolution of the animal's ancestors, or of which no precursors were present in its ancestors.[17] That is how we know humans are intelligent: our ancestors' genes would not have been able to spread better through the population if they had encoded knowledge of how to, say, make phones. Nor did our ancestors have genetically encoded precursors of such knowledge. And yet we know how to make phones, so we must have created that knowledge ourselves. When there is such evidence in an organism, we should tentatively conclude that it is a person: creative and, therefore, conscious.

Absent such evidence, I conclude that except for humans, no known organism is creative. Since free will and consciousness are epi-creative, I conclude that no organism has either. Animals cannot make decisions freely, nor can they feel pain or suffer, nor can they understand the world or have any self-awareness. With that, I consider the problem of animal welfare solved.

People's perceptions and eager anthropomorphizing not only *lead* to the misidentification of intelligence. As the case of Clever Hans shows, they can *be the reason for* the animal's behavior. The pinnacle of anthropomorphism is the demand to treat animals *humanely*.

I expect animal-rights activists to be outraged over my claims, or to reject them outright as laughable. They can get very upset. But should they not be *relieved* that animals are not conscious? It means their problem is solved, and they need not worry about animal suffering. Instead, I expect them to double down on moral grounds. They may be concerned that if I am wrong, the cost of continued animal suffering is too high, which is a modern-day version of Pascal's wager: if you believe that animals can suffer even though they cannot, no big deal. If, however, you believe that animals cannot suffer, when in reality, they do, their tremendous suffering continues. But Pascal's wager and versions thereof are bad explanations: they are easy to vary externally because they apply to absolutely anything whose potential downsides are said to be grave despite our best explanations that such downsides do not apply. By the same logic, we should treat inanimate objects with respect and dignity as well, just to be on the safe side.

Animal-rights activists may find my concern for AGI wellbeing (see chapter 8) hypocritical – presumably, someone who is not worried about the wellbeing of animals surely should not care about the wellbeing of a computer program. However, I believe the above are good explanations of why AGIs *would* be conscious and why animals are not, so if this is their criticism, they have not understood these explanations. If these explanations instead said that animals can suffer, I would be worried about the wellbeing of animals, too. But they do not say that. Intelligence and the capacity for suffering are usually offered as two separate criteria for whether or not it is okay to kill a living being. I do not think they are neatly separable like that because the capacity for suffering arises *from* intelligence. In any case, I guess what stops animal-rights activists from changing their minds is their desire to do good in the world. As I discuss in chapters 9 and 10, good intentions are a common source of evil. We should base actions on good explanations, not good intentions.

When prompted with the question of why an animal cries it out in pain when hit with a stick, I imagine animal-rights activists answering "because it really does suffer." When questioning this explanation, they would invariably reply, "how you can doubt it, seeing how it cried out?" as if this had not been considered (and refuted). And, as in the example of the dead rat from this chapter's first section – "how can you doubt the rat died of poison, seeing that it is dead?" – that is circular reasoning: we require additional claims which we can test independently.

I agree with animal-rights activists that language is not required to be able to suffer, and that language is not the ability that makes humans exceptional (Lori 2017). Nor is any other particular knowledge (except for creativity). I even argue in the next section that people need not have evolved from apes. But what is missing in the animal-welfare debate is *epistemology* to explain the inner lives of animals (or lack thereof).

Does the fact that animals are not conscious mean we need to change our relationship to our pets? No. The reason for companionship and source of affection was our anthropomorphism all along and does not need to change. The fact that you do not need to worry about your pet's wellbeing is a *positive* thing: hopefully, it makes pet ownership more manageable. Whether we then anthropomorphize our pets' emotional responses has no bearing on the absence of their inner lives, but we should not fool ourselves. Of course, we can still enjoy the company of an animal: pretending they are conscious and intelligent is amusing. We even anthropomorphize devices like automated vacuum cleaners and feel sorry for them when they let out a noise to signal they are stuck. We like to think of their behavior as asking for help. It's fun.

My denial of animal intelligence (and with it animal suffering, consciousness generally, and free will) is *not* rooted in a behavioristic interpretation of their actions. It follows from the best explanation of how animals work internally, in light of which animal behavior makes sense. Not much hinges on this view of animals, however, since the topic of the book is people-intelligence. Remember, however, that if the thought that animals do not suffer makes you uncomfortable, that discomfort – or rather, a lack thereof – is not a criterion of reality.

In truth, we should have understood the algorithmic nature of all organisms, not just animals, at the latest when the evolutionary biologist Richard Dawkins discovered that organisms are protective shields genes build around themselves (Dawkins 2016, chapter 2). Organisms are the slaves that genes use to spread through the population. They are "survival machines." (Dawkins 2016, chapter 2) As such, genes need to have maximum possible control over their bodies. All living organisms, from single-celled organisms to mighty humpback whales, are machines whose physiology *and behavior* are programmed by their genes. (Human behavior is an exception; I explain why in the next section.) The sophistication of such behavior is the result of biological natural selection, not creativity. Biological evolution is the creative phase, which is interspersed with the organisms' lifetimes, the perspiration phases. As the ethologist Konrad Lorenz said: "what is a priori for the individual is a posteriori for the species" – that is, a priori/posteriori in terms of *knowledge*, not psychologically. The physiology (hardware) and behaviors (software) are genetically given.

The algorithmic nature of even rather sophisticated animal behavior is easily understood when the lack of error correction and the lack of creativity are made apparent. For example, there are dozens (if not hundreds) of videos (YouTube 2015) available online of dogs that, when held above a surface of water, such as a bathtub or a swimming pool, start kicking their legs in a swimming motion, without even touching the water. Swimming is somewhat sophisticated, and its presence explained easily: natural selection has endowed dogs with an algorithm to recognize water and kick off the swimming motion because dogs that exhibit this behavior spread their genes more than those that do not.

However, it is intuitively clear that no intelligence is at work here. Why? Because an *intelligent* being would have noticed *errors* in its behavior and corrected them. If one of these dogs were intelligent, it would know that it is being held *above* water, and therefore need not swim. The dogs in these videos do not even realize they remain stationary because they are being held. In other words, to be intelligent, they should *expect to move* while swimming, and have the *desire* to understand the problem when this expectation is disappointed. There appears to be no such desire for error correction in dogs or other animals. In some videos, the pools are not even filled with water. One dog is merely being held over a sprinkler and does the swimming motion anyway. One anonymous commenter on the referenced video points out that the condition (he calls it "instinct") in dogs that kicks off their swimming algorithm seems to be: if not touching the ground and sensing water-like surroundings, start swimming. ("Water-like" may simply mean "cold" – a video of a dog being held above ground in front of an air conditioner shows it kicking its legs as well. (Giphy, n.d.)) The rule is simple, and the swimming algorithm is admittedly sophisticated, but that is all there is to it.

Therefore, the sophistication of animal behavior (which comes in degrees) is no evidence of intelligence (which is a binary matter) since even the most impressively sophisticated behavior can be the result of incremental honing over millions of years of biological evolution. Only the ability to correct errors can be evidence of intelligence.

That being said, we do encounter a problem with this theory. Consider the following situation. A well-trained dog will perform tricks on command fairly reliably. What happens when you tell a dog to repeat the same trick over and over again in rapid succession? Depending on how well-trained the dog is, it may repeat the trick a dozen times, maybe even more often than that. But eventually, it will stop. Likewise, I guess that dogs that are being held above water do not kick their legs *forever*. They do not keep kicking until they faint from exhaustion. I guess they stop long before that.

So, is there error correction happening at runtime after all? I do not think so. A better explanation is that there is a *meta-algorithm* or "failsafe" (Temple 2019a) at work that interrupts the dog's behavior after identifying that it got stuck in a loop. This meta-algorithm plays a crucial role in all animals, including people. It would make sense for such a meta-algorithm to evolve, for, whenever an animal got stuck in a loop, it almost certainly died. So, evolution "considered" getting stuck in a loop an error because it led to some genes' death. I suspect that many organisms died in such a manner until chance genetic mutations happened to solve this problem. Thus, biological evolution solved the problem in terms of interrupting the loop and encoded the solution in the dog's genes. We do not need creative error detection or correction happening at runtime to explain this.

I assume that most, if not all, organisms have such a meta-algorithm. It performs predetermined actions (functions) encoded in a policy based on external circumstances (for example, "if predator visible, run away"), and internal states ("if tired, rest"), or both ("if food visible and hungry, move toward it and eat it"). In other words, a genetically encoded meta-algorithm, itself a function, continuously monitors the internal state and external stimuli and wraps other genetically encoded functions, which it invokes in response to the monitored results. More importantly, the meta-algorithm is a proto error-correction facility. It finds and corrects errors (such as getting stuck in a loop (Temple 2019a) or not fleeing from a predator while eating) according to genetically predetermined mechanisms. However, in this form, it cannot detect or correct *new* errors that occur at runtime (such as attempting to swim while being held above water) and, therefore, cannot be considered creative. In animals, the meta-algorithm solves the *genes'* problems, not the individual's.

Nonetheless, this meta-algorithm played a vital role in the evolution of people. And the reason why animals are not intelligent follows from the fact that they miss a simple genetic mutation that we all have inherited from a single human ancestor.

The Evolution of Creativity

If consciousness and free will really are epi-creative, then we can explain their evolution through the evolution of creativity. They would have come along for the ride at the point creativity evolved. But here a problem awaits us: if no organism other than humans is creative, how did creativity evolve? One potential solution is that it did not and that the entire chain of humanity's ancestry was also creative. That cannot be so, for, according to our best explanations of life on Earth, all organisms that are alive today, including humans, stem from a few ancestors, or even a single ancestor. This ancestor lived in the primordial soup, the hypothetical conditions present on Earth around four billion years ago, in which the first self-replicating molecules formed. It is doubtful that those molecules were creative. Indeed, judging by evolutionary timescales, even fairly recent ancestors of ours were not creative. So instead, I must explain the evolution of creativity from non-creative ancestors. I will also call into memory a few additional problems we have run into so far.

At the end of this chapter's first section, we faced three problems. First, Popper's theory of knowledge is an evolutionary theory. But in its current form, it lacks the notion of a *replicator*, without which evolution cannot occur. Knowledge evolves according to the same underlying logic everywhere in the universe. This logic depends on the neo-Darwinian notion of *replication* (Dawkins 2016, chapter 13). Evolution explains the creation of knowledge through the "non-random differential reproduction" of alleles (Dawkins 2016, foreword to first edition) – slight variations of a replicator – in a pool of replicators. While we have an evolutionary theory of *memes* replicating *across* minds (Dawkins 2016, chapter 11), we do not yet have a theory of knowledge replicating and evolving *within a single mind*. Before a piece of knowledge can become a meme by spreading to other minds, it must first evolve within one mind.[18] And a complex meme cannot spread to a mind that is not creative. In both cases, we first need replicators in a single mind (among other things).

Second, we have had to confront multiple infinite regresses in our current understanding of creativity: for example, a mind creatively conjectures problems and criteria for criticism. But since creativity, in turn, rests upon problem detection and criticism, we have an infinite regress.

Third, we do not yet understand the mechanism that gives rise to conjecture: if my discussion of consciousness is correct, conjecture happens subconsciously, so we cannot learn about it through introspection. We know it occurs, but we do not yet know how.

I believe that all four problems, which include the evolution of creativity from non-creative ancestors, have the same solution. And it is the story of the evolution of creativity that allows me to illustrate how creativity may work, and in turn, how one might build AGI.

All organisms exhibit behaviors which must be encoded genetically as functions. Even the simplest organisms know how to, say, swim in a particular direction and turn around when they encounter an obstacle. If the knowledge causing these behaviors were not stored in the organism somehow, it could not enact them. I call the set of functions that are available to an organism its *function pool*. (This is similar to, but not to be confused with, the gene pool, for a function pool lives within a *single* organism, whereas the gene pool is spread across *all* organisms.) As I discussed, each organism contains a meta-algorithm (itself a function) that oversees the function pool and chooses which functions to invoke through an interplay with the organism's policy and based on the organism's internal state and external stimuli. It further contains genetically predetermined detection and correction of errors according to likewise genetically given criteria. Such criteria are also functions and part of the function pool.

I suspect that even the most primitive of our ancestors contained these features in a basic form. Their meta-algorithm consisted of a loop that continuously checked the internal state and external stimuli and invoked or interrupted running functions accordingly. The ability to interrupt function execution, in particular, was of great value: if a predator ever showed up while the organism was busy eating, the organism needed to be able to stop eating and flee. As discussed, the meta-algorithm further contained safeguards against getting stuck in a loop. (This was merely the kind of mark biological evolution had left on the meta-algorithm in terms of predefined error correction; no errors could yet be detected creatively, outside some predefined means.) The only loop permitted to be never-ending was presumably the meta-algorithm itself, for there is no meta-meta-algorithm to watch over the meta-algorithm.

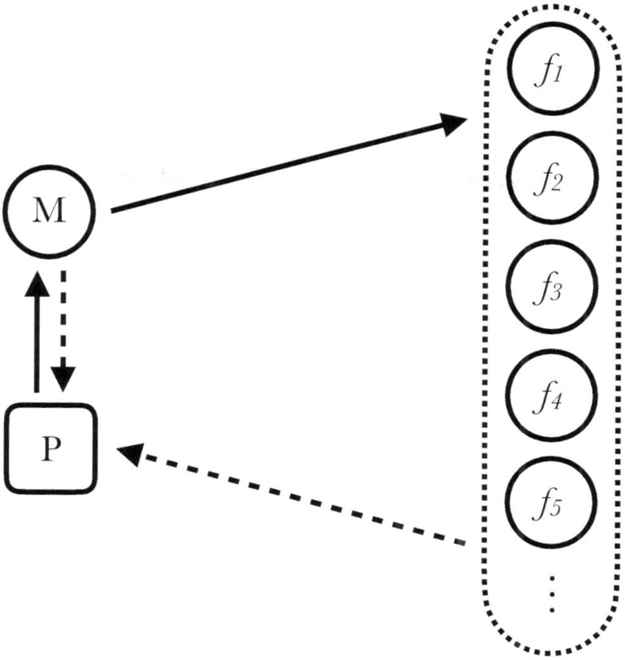

A simplistic view of a non-creative mind. The meta-algorithm M continuously invokes functions f from the function pool on the right based on a policy P. Both M and any f can update P. All of M, P, and f are genetically encoded. The function pool is largely static.

Since the meta-algorithm, the policy, and the function pool were encoded genetically, they were subject to biological evolution. Their instructions did not change within a single organism's lifetime but were subject to occasional mutation whenever an organism reproduced. As a result, any of them could have been slightly different in the organism's offspring. Since biological evolution had already adapted any given organism to its environment, most changes would have been disadvantageous. However, an occasional advantageous mutation led to an improvement in either the organism's meta-algorithm, policy, or function pool. Whenever that happened, the organism's offspring had a better ability to spread its genes through the population. That is how, over time, the implementations of all three improved. This improvement led to the sophisticated behavior we see in organisms around us today, from the honey bee's intricate dance over horses with facial-feature-recognition algorithms to complex smell, facial, and predator-recognition algorithms in dogs and other animals.

It is notable that still today, an organism's function pool consists of functions determining behavior, but not the organism's physiology. Those two types of functions seem strictly separated. While genes execute careful control over the organism's early development, there do not seem to be functions the organism can invoke to change its physiology once it has matured (with some exceptions, such as chameleons). But of course, an organism's body has interfaces its functions can invoke in order to, say, move around. When I speak of functions here, I mean functions in the behavioral sense (and later on, as we will see, in the sense of explanatory knowledge).

In any case, in some organisms, functions in the function pool triggered a certain kind of side effect. It was the addition of functions to the function pool. This way, a function that was present only as part of another function became invokable by the meta-algorithm as a standalone function. Until then, it was dormant, and its addition to the function pool can be considered a kind of maturing. We do not know much about the meta-algorithm's evolutionary history. It may have biologically evolved pruning mechanisms in response to the pool getting too large relative to the organism's memory capacity, occasionally clearing it of those functions that had not been used in a while, or which, according to some predefined criterion, were not useful. This process went on inside organisms for millions of years.

But then something different happened. It was similar to the autocatalytic origin of life – only this time, it took place within the mind of a single individual. Within one such organism – an animal, presumably an ape that was our ancestor – a genetic mutation was present that led to a new kind of function. In addition to what it normally did, it added some of its building blocks back into the function pool. It was not yet targeted enough to be considered a replicator, since it may also have added other building blocks that were not related to its original functionality. However, it was the first hint of a replicator (a "proto-replicator").

Whenever this function was invoked, it would add more of its building blocks to the function pool, some of which implemented its functionality. Throughout the individual organism's lifetime, those building blocks which, taken together, best caused their production relative to their variants, would spread through the function pool, making it more likely for the meta-algorithm to invoke them.

Gradually, evolution produced replicators that copied themselves in large numbers, quickly, and reliably – as well as ones that were able to evade the meta-algorithm's pruning mechanism. Only this time, it was not *biological* evolution: it was evolution occurring *within this single ancestor's mind*. In this context, it is not the gene that replicates, but the function, and I refer to it as *functional evolution*[19] to disambiguate it from biological evolution. Though the underlying mechanisms of non-random differential reproduction of alleles are the same, the actors and selection mechanisms are different.

That this event was similar to the origin of life in the biosphere (see also chapter 2) should not surprise us. It was the origin of life that kicked off the creation of genetic knowledge, and so it would make sense for the beginning of the creation of human knowledge to be similar to the origin of life. The difference is that the origin of life had to happen only once, whereas the origin of human knowledge happens in each of us at the beginning of our lives.

After that mutation occurred – we are still picturing this single ancestor's mind – whenever the meta-algorithm invoked a replicator, this replicator created a copy of itself with high but imperfect fidelity and added it to the function pool. Throughout the ancestor's lifetime, large quantities of that same function and slight variations thereof were present in his function pool and available for the meta-algorithm to invoke. The emergence of slight functional variants is similar to genetic drift (Popper 1977) – perhaps deserving of the name "functional drift" – in the sense that over time, even in the absence of any selective pressure, the frequency of alleles in the function pool changes. Alleles are in fierce competition with each other (Dawkins 2016, chapter 3).

Functions competed[20] over the only two limited resources that apply to the mind, that is, memory and processing power, as well as over the meta-algorithm's "attention" so that it would invoke them, allowing them to replicate further. A good replicator causes the meta-algorithm to invoke it. The moment imperfect replication of functions was coupled with competition between them, natural selection kicked in, and the seemingly random functional drift started having the appearance of being directed toward a purpose. For the first time, evolution was present in a mind, at runtime, which is a prerequisite for creativity. Neo-Darwinian evolution had been instantiated in a mind.

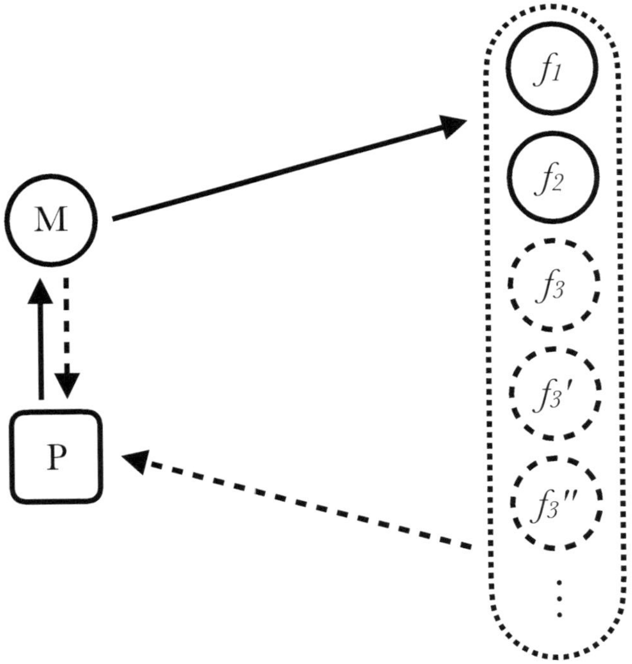

Through a genetic mutation, function f_3 has turned into a replicator. The function pool becomes dynamic. Due to copying errors, functions emerge that are not genetically encoded.

Most replication merely led to the copying of some inborn behavior. Since functional evolution, like all kinds of evolution, favors high copying fidelity (Dawkins 2016, chapter 2), mutations to replicating functions would have been rare. At the beginning of that individual's lifetime, most of them would have been only slight variations of the single built-in function that replicated. Should a mutation have occurred that would have lead to detrimental behavior on the level of the individual, its chance of being invoked would have been small.

Now, what is advantageous from the function's perspective is not necessarily advantageous to the individual organism, let alone the gene, and enacting the function in the form of behavior may well lead to the organism's death. (This explains how ideas can lead to the death of people, and with them, their DNA, despite creativity being genetically encoded and genes trying to spread through the population.) Nonetheless, even in these early stages of the biological evolution of creativity, the meta-algorithm already had built-in ways to detect and interrupt behaviors that might be detrimental to its host, such as getting stuck in loops. Therefore, the individual's behavior was not suddenly dominated by the replicating function, enacting that function over and over again, just because it was much more numerous than any given built-in function. The meta-algorithm would have identified that as being stuck in a loop and terminated that process. This "failsafe" (Temple 2019a) also applied to function alleles, which may not have led to sufficiently different behavior not to be identified as being stuck in a loop, too. They would not have gone unnoticed by the meta-algorithm.

Initially, the vast majority of replicators at any given time were either exact copies or alleles. Nonetheless, as the evolution of functions continued within this individual's mind, function alleles were invoked from time to time, mutations added up, and completely new functions eventually emerged, leading to novel behavior when invoked. If, on average, these new behaviors led to a slightly increased chance of reproduction, biological evolution would have spread creativity through the population. Provided the meta-algorithm had sufficient reach in its ability to detect any apparent blunders, completely new functions and behaviors emerged relatively safely. The biological evolution of a replicator within this ancestor's mind should not have led to any disadvantage. So from this perspective alone, this single ancestor's chances of having offspring should not have changed significantly for the worse, allowing creativity to spread through the population. (Not only did creativity not *hurt*, but it led to an *advantage*, as I will explain shortly.)

Functional evolution happens much more rapidly than biological evolution because it does not need to "wait" for the organism to have offspring. Genetically encoded functions had to compete with those created at runtime, but, since they could not replicate, they were quickly outnumbered over time and rarely invoked by the meta-algorithm as a matter of probability as soon as new functions emerged.

This phenomenon explains why genetic determinism, while applicable to animals and other organisms, does not apply to humans, and why it is so easy for humans to ignore the few built-in instructions they have. Genes do not influence human behavior very much. They provide inborn knowledge, which can be overwritten at runtime. It is subject to improvement, like all our ideas. This overwritability explains the possibility of the evolution of ideas such as birth control and homosexuality, which the gene should consider its worst enemy. The vast majority of knowledge any person has he creates during his lifetime – i.e., at runtime. Some types of inborn knowledge can be harder to overwrite than others: it may be more challenging to overcome a fear of heights than to, say, learn to enjoy the smell of spoiled milk, and it may be almost impossible to suppress certain reflexes. Nonetheless, people fast in defiance of hunger and jump out of airplanes in defiance of their fear of heights – even to have fun (Deutsch 2015, 1:11:47). Evolutionary psychology (which seeks to explain human behavior in terms of *biological* evolution rather than *functional* evolution, and which says that our genes determine our minds and our psychological makeup) ceases being relevant shortly after birth at the latest.

As creativity spread through the population of our ancestors, it saw the occasional genetic mutation of built-in functions, usually disadvantageous or harmful. But now creativity could make up for these mutations by functionally evolving replacements at runtime. Since genetic mutations, like all mutations, are often detrimental, *biological evolution favored the improvement of creativity at the same frequency at which detrimental genetic mutations occur naturally*. Every such mutation increased the selection pressure in favor of creativity, so from that point on, creativity spread rapidly. Any mutations that usually led to an individual's death and that creativity could now offset allowed the mutated individual's genes to spread. In other words, creativity improved its genes' fecundity, and, thereby, their ability to spread through the gene pool.

There would have been no need to favor genetic fixes to built-in functions damaged by mutations. So they deteriorated further through the generations, with creativity taking up ever more work that used to be done by genes. Since the deteriorating genes had once been hard to vary while fulfilling their purpose, any adequate, functionally-evolved replacement was likewise hard to vary and, therefore, had sticking power. Over time, biological evolution genetically kept only those built-in functions that were too complex to recreate quickly enough at runtime and that were needed early enough in life. That is why today, very few human behaviors and ideas are still inborn; babies are born "half baked," with almost no faculties present except for a few instincts such as sucking on the mother's breast – and creativity. In other animals, the entire repertoire of available behaviors seems to be present at birth – though some cannot express it immediately because their physiology needs to mature first, and because some functions are added to their function pool only later in life.

Since creativity increased the fecundity of its holders' genes, biological evolution would have favored mutations improving the creative algorithm, including the brain's memory capacity and processing power. Such improvements would also have included the gradual replacement of proto-replicators with actual replicators so that a baby's mind did not need to evolve replicating functions from proto-replicators. This way, function replication started as soon as the child's meta-algorithm came online and invoked the replicator for the first time (possibly while still in the mother's womb!). Presumably, natural selection in biological evolution favored meta-algorithms that happened to invoke the replicator early on in order to kick off the creative process as soon as possible.

Other factors, such as the function replicator's mutation rate and the interplay between the meta-algorithm and the evolving functions, would have improved as well. Through small, incremental changes, biological evolution further cemented the parameters of functional evolution while replacing ever more inborn functionalities.

However, impressive as their new abilities may have been, this alone would not have made our ancestors *people*. We could perhaps consider them *proto-creative* since functional evolution occurred within their minds, but they would have still lacked the ability to detect *new* errors creatively at runtime. Since their meta-algorithms only evolved biologically, from individual to individual over generations, only biological evolution could have detected new errors, at the slow speed at which it occurs. Error correction at runtime is a crucial ingredient of creativity. The meta-algorithm may have pruned functions that would have gotten the individual killed. However, contradictory functions, though not lethal, must nonetheless have evolved functionally in our ancestors' minds, which sometimes led to inconsistent and even erroneous behavior.

So from the perspective of biological evolution, there was a need for the detection of new errors at runtime. Now that *something* other than the automatic execution of predefined functions was happening at runtime – namely the evolution of functions – there was a real opportunity to functionally evolve the detection of new errors and hook it into the already running process. For, should any gene happen to bestow its proto-creative holder with this ability, it would have a tremendous advantage at spreading through the population.

I guess that in the presence of the meta-algorithm functions evolved specific roles for themselves. Functions had to "convince" the meta-algorithm to invoke them in order to replicate, so they needed to fulfill whatever purpose the meta-algorithm sought to meet. Since one of the meta-algorithm's primary responsibilities is to look for errors and correct them, it would have invoked functions that evolved to assist it in this task more often. And, since they replicated whenever invoked, they would thus have "ensured" their spreading through the population of functions.

In other words, if some functions happened to provide criteria the meta-algorithm could use to criticize other functions – that is, detect *problems* – as well as criteria to determine whether something should be considered a solution to a given problem, the methods for error correction themselves became subject to change at runtime. That is how functions evolve the roles of problem detectors and critics. It explains how criteria for selection (criticism) and function invocation (taking action) can themselves be subject to change at runtime, which they seem to be in humans. So during an individual's lifetime, the meta-algorithm's policy itself became subject to changes that were not predefined genetically, and the meta-algorithm slowly turned into a more sophisticated error-correction mechanism.

This change may well have occurred in that first proto-creative ancestor, making him *fully creative*. Since consciousness seems to be linked to error correction, this change may explain the evolution of consciousness. I guess it evolved in order to perform more sophisticated selections in the function pool. It may have co-evolved with function replication in our minds: consciousness led to better function selection, which in turn led to more awareness of one's knowledge and improved knowledge of how to use it, which in turn lead to even better function selection, and so forth.

I guess from this point on, through the same selective pressures that drove the biological evolution of proto-creativity, biological evolution favored those configurations of the creative algorithm in which functions evolved these roles of problem detectors and critics as quickly as possible. This conjecture solves the various infinite regresses in Popper's epistemology, such as those behind problem detection and criticism: *problems, criticisms, and solutions are themselves adaptations whose presence we need to explain in terms of functional evolution*. They are evolved reproductive strategies of functions, and they emerge through the functional evolution that occurs in the mind. And the problem of the mechanisms behind conjecture is solved in terms of chance mutations of replicating functions.

Darwin was troubled that he could not explain variation in biological evolution (Popper 1977). He could not explain it because he did not know about the gene generally, nor that it is the unit of selection. But we do know the unit of selection in the functional evolution that takes place in people's minds: the self-replicating function.

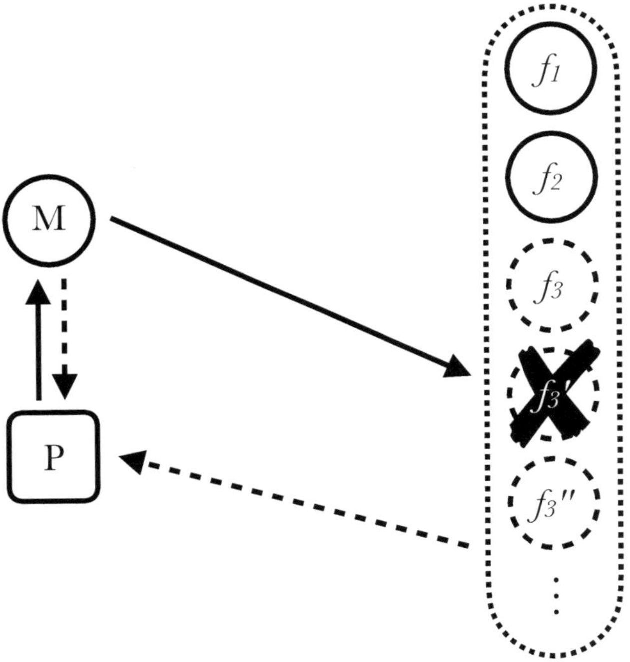

The meta-algorithm M can detect problems and prune functions according to functionally evolved error-correction criteria. The mind becomes fully creative.

The creative algorithm is complete once error detection and correction occur at runtime according to newly conjectured criteria. Our first ancestor to have had this ability may well have been the first creative and conscious animal ever, and with that, the first person ever.

I guess that a person's meta-algorithm is the conscious "I" we each feel we have; it feels like a control center because that is what it is. Evolution of any kind is an interplay between bottom-up and top-down causes. Replication and mutations happen at the level of the unit of selection – the function – and thereby affect the frequency of function alleles (bottom-up), while the meta-algorithm, through its selective properties, weeds out some functions (top-down). This top-down causation is the reason the "I" is – or is at least associated with or a property of – the meta-algorithm. It is why we feel (correctly) that we make decisions and observe our mental state top-down, as well as command our bodies like puppets on strings guided from a command center – again, top-down – to affect changes in the world.

It is the possibility of top-down causation generally that reductionism denies, so I guess that no reductionist theory can explain consciousness. The theory of biological evolution was at heart a philosophical breakthrough, for it represented an alternative approach to how we can formulate scientific theories. Until Darwin thought of the theory of evolution, the only mode of explanation was that of initial conditions and laws of motion predicting some outcome. But rather than being given initial conditions and laws of motion to predict with certainty what *will* happen, the theory of evolution does not predict any particular organism. "There are no Darwinian laws of evolution." (Popper 1983, 267) Evolution merely explains *how* an organism of any kind *can*, in principle, arise in nature.

By the time creativity finally evolved, our ancestors already had simple memes, which they copied by literally imitating behavior according to genetically built-in methods (Deutsch 2012a, 405-409). These simple memes also contained knowledge that genes did not have to provide anymore, such as "choice of mate, and methods of eating, fighting," (Deutsch 2012a, 414-415) et cetera, so co-evolution between memes and genes was already happening. As a result, even before creativity came on the scene, biological evolution had been exerting selective pressure favoring genetic mutations that increased storage capacity in brains, to non-creatively copy and enact ever more complex memes.

(This increased storage capacity later became the fertile ground required for functions to replicate within a mind.)

However, the content of a meme is inherently limited if it can only be copied by directly imitating behavior – in other words, if it is implemented behavioristically. People memes, that is, explanatory memes, are *rule-based* (Deutsch 2012a, 417). They give rise to behavior in the meme's holder when invoked by the holder's meta-algorithm. If the rules themselves are hidden away in people's minds and only expressed through behavior, how can these rules replicate *across* minds? The meme's "recipient" can only do this by *explaining* the "sender's" behavior (Deutsch 2012a, 410), which he can only do through creativity.

Creativity enabled its holders to transform behavioristic meme implementations into rule-based ones. It allowed them to replicate ever more complex behaviors by explaining them, and there would automatically have been selection pressure to spread and improve creativity (Deutsch 2012a, 416).

To explain it in terms of the example of addition from chapter 3: up until the evolution of creativity, people could only replicate memes that had the structure of addition tables – i.e., behavioristic implementations of addition – by literally copying each entry in the table. Once creativity evolved, they could *explain* such tables, that is, recreate the rules that produce them. Those rules require less memory to store, yet have much further reach: they go *beyond* the already made observations. In some cases, they reach to infinity. Explaining is also much faster than copying each observed entry manually.

For example, the meme of addition would have been stored in a non-creative ancestor's mind like our instrumentalist implementation from chapter 3:

```
(defn plus-instrumentalist [a b]
  (cond
    (and (= a 1) (= b 1))
    2
    (and (= a 1) (= b 2))
    3
    ; ...
    (and (= a 2) (= b 1))
    3
    (and (= a 2) (= b 2))
    4
    ; ... and so on up to some limit
    ))
```

A creative ancestor would then have been able to copy it by *explaining* it, thereby turning it into something like the good implementation of addition from the same chapter:

```
(defn plus-rule-based [a b]
  (if (zero? b)
    a
    (recur (inc a) (dec b))))
```

(I do not mean to imply that our pre-human or even early human ancestors knew how to perform addition; they did not. It is merely an example of the structure of behavioristic functions versus rule-based functions. A more realistic example would be how to make fire, where each entry in the table corresponds to an observed part-action of making fire, but the same fundamental change from behavioristic to rule-based implementation applies.) Before the evolution of creativity, memes' structures were like those of addition tables; afterward, they were rule-based, that is, explanatory, function implementations.

Over time, this gave rise to the evolution of complex memes such as today's languages. And it is not only new *behavior* that started to evolve functionally. More importantly, functions began to evolve within minds and across minds that did not trigger any particular behavior or any other side effects, but instead were explanations – of anything at all. Since functions have computational universality, creativity automatically conferred the ability to create *all kinds* of knowledge: not only scientific, but philosophical, artistic, moral, emotional, mathematical, and otherwise. For the first time in the history of life on Earth, there was a new kind of program – a person – that could understand the world. For, whenever a functional mutation led to a function that happened to replicate, say, a physical process, the holder had then understood that process and had thereby created a bit of scientific knowledge.

By the way, the set of functions a creative mind can, in principle, produce includes all possible functions that cause side effects. As I have said, such functions, when invoked, cause a physical change in the world. The set of all possible side effects is infinitely big and ranges from something as mundane as printing text on a screen to something as momentous as building a structure in space that can move a black hole, and it includes other side effects even more magnificent that we have not conjectured yet. Deutsch says that people are not only universal explainers, but also universal constructors: they can transform "*anything into anything* that the laws of nature allow." (2012a, 59) Being universal explainers "makes them universal constructors as well." (2012a, 146) I am no expert on Deutsch's Constructor Theory (2013), but I guess the fact that being a universal explainer also endows the abilities of a universal constructor is due to the enormous reach of side effects. It is people's universal ability to evolve functions with arbitrary side effects that makes them universal constructors. Differently put: being a universal explainer allows you to explain (and, therefore, cause) any possible physical transformation.

In any case, I believe that the single genetic mutation I presented above is how creativity evolved. It is the reason our species exists.

If my theory is true, the creative algorithm is entirely inborn. Because memes and genes had already been coevolving before creativity came on the scene, with memes taking up ever more complex functionalities that were originally genetic, Deutsch speculates that the creative algorithm is only partly genetic and also partly memetic (2012a, 414-415). While I agree that memes played a significant role in our evolution, I disagree that the creative algorithm is not entirely inborn. For, my explanation of the genetic jump to creativity likewise incorporates the important role of memes, but explains creativity as something entirely inborn and kicked off by genes only. And, as far as I am aware, that explanation is the best available one (perhaps the only non-refuted one) not only of what factors enabled and drove the evolution of creativity but of *how* creativity evolved, in detail. Therefore, following Deutsch's criterion for reality (2012a, 23), I propose that my genetic-jump explanation shows what really happened in our evolutionary history and that newborn people are creative without first acquiring any memes.

Nonetheless, I agree with Deutsch that explanations of the evolution of our species in terms of upright posture, bigger brains, diet, changes in available resources, or other environmental circumstances, opposable thumbs, et cetera are all false (Deutsch 2012a, 415-416). The fact that it was a kind of ape that evolved to become a person is incidental. What was needed was a change in *software*. It could have been any animal whose meta-algorithm had enough reach to catch most fatal errors, as long as a genetic mutation occurred that introduced a self-replicating function. For all we know, people could have evolved from honey bees; new people still might. Any other animal today can, in principle, become a person, too, as long its genes contain the requisite mutation that introduces a self-replicating function in its mind.

I guess this is, roughly, how creativity works today. Unfortunately, creativity by itself does not make its holder rational (though it does confer the *ability* to be rational). The knowledge of how to be rational, that is, to look for good explanations and actively correct errors, must itself evolve functionally (or memetically) from much more primitive criteria for problem solving.

Functions evolve specific roles that fulfill purposes (such as criteria for problem detection and correction) to get themselves invoked by the meta-algorithm and, thereby, to replicate. They are "dynamic" or "rational" functions because they help their holder detect and solve problems. However, there is also a counteracting force present in any individual's function pool. It consists of functions that evolve to *evade* detection and criticism. Once invoked, they divert the meta-algorithm's resources into maintaining some hangup, so they can continue replicating without their introduction of errors being noticed. They are "static" or "anti-rational" functions. (I will say more about both kinds of functions, which are based on Deutsch's work on dynamic and static memes (2012a, 388-390), in chapter 9.) It is due to static functions that for most of human history, people did not use their creativity to innovate, but to faithfully replicate and enact behavior, customs, and traditions (see also Deutsch 2012a, 402).

Lastly, we should expect to find functional replication strategies similar to those we have found in genes. For example, some functions may not include the instructions for their replication anymore because they "cooperate" with other functions in the mind that will replicate them whenever *they* are invoked.

The relationship between the evolving function pool and the meta-algorithm is an interplay between "something haphazard" and a "selective control" (Popper 1983, 231-232). It is only under these conditions that knowledge can grow. We do not yet know much about how the meta-algorithm works, but here we have an advantage: we roughly understand how the evolutionary part works. We also know that the meta-algorithm need not have changed much in order to accommodate creativity once functions started replicating. Therefore, it should be possible to perform tests on animals, our close relatives in particular, in order to reverse-engineer how their meta-algorithms work. Their functionality may shed light on how ours works. At the very least, as I have suggested, we know the following things about it: it looks for errors according to given criteria and invokes functions as fixes for those errors according to likewise given criteria. Presumably, it operates in a loop and invokes functions based on the internal state of the organism and external stimuli. It contains knowledge (functions) of how to interpret both and map them onto the appropriate functions to invoke. All of this together constitutes the meta-algorithm's policy.

I guess whenever the meta-algorithm invokes a function from the function pool to find a flaw with it, that constitutes a *thought* (though an explanation of the quale of seeing it in front of the mind's eye continues to elude us). Whenever that function invokes another in its turn, which can occur in a deeply nested fashion until a function is reached that terminates without invoking another function, then such a string of function invocations constitutes a *train of thought.* (However, note that most function invocations seem to happen subconsciously – i.e., they do not seem to occur for the purpose of creative error detection or correction.)

I further guess that the meta-algorithm can interrupt single thoughts and trains of thought, and that, once a single or train of thought terminates, it looks for other functions to run.[21] This way, the organism rarely idles. Multitudes of copies of a single function are not perceived as such, because the mind only runs one function at a time, i.e., it only experiences/thinks of one idea at a time. Even if the brain has multiple parallel processors, the number of processors – and with it, the number of simultaneous thoughts – is presumably low compared to the size of its function pool. Therefore, it would make sense for the meta-algorithm to invoke functions in such a way that no function runs several times at once.

This theory also (roughly) explains the origin of thoughts that just pop into consciousness: they are functions that evolved through a long chain of variation and criticism, invoked by the meta-algorithm to find or correct errors. I suspect we can reverse-engineer parts of the meta-algorithm through introspection. You can see it at work when you close your eyes, quiet your mind, and observe thoughts appearing out of nowhere. There must be a logic to how this happens: thoughts of some type may be invoked in rapid succession, or depending on the time of day, or based on other factors.

We have also solved the more general problem with Popperian epistemology I discussed earlier concerning its lack of replicators. While it has significant parallels to evolution (conjecture is variation, and criticism is selection), evolution is a theory about replicators (Dawkins 2016, chapter 2). Since, as I have argued, evolution is the only explanation we have of a process that creates knowledge, and since evolution is a theory about replicators, we needed to introduce replicators into Popperian epistemology. The individual function is the replicator and the unit of selection. What is it a function replicator is "trying" to do in a mind? It "wants" to spread through the population of that mind's function pool. Every individual function "struggles" to spread. (Though, of course, it is not "aware" that it is struggling (Dawkins 2016, chapter 2); speaking of functions as if they were conscious actors is merely convenient language. The only conscious function seems to be the meta-algorithm.)

Over time, the mind's function pool will consist of those functions that can replicate fairly faithfully compared to their rivals, can survive longer than their rivals, and can replicate faster than their rivals. In other words, the evolution of functions favors copying fidelity, longevity, and fecundity, just like biological evolution and memetic evolution do (Dawkins 2016, chapter 2 and chapter 11).

Functions compete for memory and processing power. Both are finite resources in a single mind, so it cannot support an infinite number of functions. Since they can only replicate when invoked, functions also compete for the meta-algorithm's "attention." As I said, they do so by evolving specific reproductive strategies: problem theories (functions alleging a conflict between other functions), criteria for criticism, and solutions. They are all conjectures because they are the result of chance variation through imperfect replication. Since the meta-algorithm prunes those functions that do not meet some criteria, function implementations less skilled at replicating die off, and the function pool is dominated by those functions which are better replicators. It is this interplay between the meta-algorithm and the evolving function pool that constitutes creativity.

It will be hard not to think of an idea of which there are many copies in one's mind: those ideas that are good at replicating will dominate one's thoughts. Why do we not slow down looking for solutions as functions replicate, having to take into account ever more conjectures as potential solutions, and comparing them against ever more criteria? Because a good explanation will take the place of many bad ones – it has a unifying character (Popper 1983, 369) and will, therefore, free up memory and processing power. It spreads through the population of functions at the expense of its rivals. A person with good explanations will have more available memory and processing power; he can also be more creative as his function pool is not yet taken up by bad explanations.

This theory of the mind has sufficient reach to solve other problems that it was not originally designed to solve. For example, the non-random differential reproduction of functions explains why it can be challenging to break a habit: many copies of the original function that codes for the behavior may still be around in one's mind and invoked frequently. To break a bad habit, one first needs to solve the problem these functions cause. That way, the meta-algorithm can begin pruning some of those copies, and they become fewer in number. And even once one has solved the problem, old copies of the original, mistaken function may still be around, lurking in one's mind, waiting to be invoked so they can spread again. It is similar to how bad ideas (e.g., fascism) can stick around in some subgroups of people even though better ideas in society at large have conceptually replaced them. The original idea may have influenced the policy in such a way that in the future, given the right circumstances, the meta-algorithm will invoke one of the idea's dormant copies, and the replication process starts over.

The theory further offers a new explanation of *memory*. (Here I mean "memory" as in "recalling things," not as in "storage space.") Memory is also a phenomenon that emerges from the non-random differential reproduction of ideas: when many copies of an idea are present, the memory of that idea is easy to recall because out of all the ideas in the function pool, the one the mind "looks for" will be selected with higher likelihood (just like one is more likely to, say, hit three quarters of a dartboard than the remaining quarter). Conversely, when only a few copies of an idea are present, it is less well represented in the population and, therefore, harder to find. Recalling it, that is, finding it will take longer. When only variants of an idea are present, and the original has gone extinct, the variants need to be used to reconstruct the original.

The need for such reconstruction may be what we experience when something is on the "tip of our tongue"; the best found variant so far is *similar* to the original, but *not quite the same*, and it is not yet clear how to construct the original from the variant. In many cases, reconstruction of the original idea from a variant is faulty, which explains why memories are often unreliable, as exemplified by notoriously unreliable witness statements in court. Non-random differential reproduction of ideas also explains why some memories fade faster than others: it is because some ideas in the mind's function pool go extinct faster than others.

In other words, the theory explains why older memories are harder to recall, why older memories are fewer in number than more recent memories, and why older memories are less reliable than recent memories.

So much for the retrieval of memories. But the theory can tell us about *committing* something to memory, too. When we create a new idea, especially if it is a good solution to a problem, we then think of it intently, excitedly, and invoke it many times to investigate its many claims, for the idea is rich with consequences. We thereby end up inadvertently copying it every time, because whenever we invoke it, it copies itself. The same happens when we practice a skill through repetition. It is not the repetition itself that constitutes learning. Instead, during repetition, a sea of copies of the same function is created, which copies we submit to criticism. Some are slight variants that may be better, which we then invoke in turn, and we keep iterating over this process until functions end up converging toward a solution that satisfies the applied criteria. At that point, some of the worse replicas are still around in the function pool, and we may occasionally invoke one of them, thereby repeating a past mistake. But if over time we occasionally invoke both the chosen criteria for criticism as well as the function itself, it helps all of them spread through the function pool again, thereby staying fresh in memory. When only very few functions that are erroneous vis-a-vis a particular skill inhabit your function pool, you can claim to have mastered that skill. Do not rest on your laurels, though: if you do not keep critically assessing your skill, functions will have the freedom to mutate away from the skillful state and your ability will deteriorate. As Popper explained, knowledge is not reliable (1983, 22). The reason is that functions replicate imperfectly.

Thus, the theory contains learning through repetition as an approximation. It explains practicing something through repetition in terms of ensuring that more of those variations are created that correspond to the desired state than those that do not. And with that, it explains the apparent goal-directedness of problem solving, including the apparent immediacy and goal-directedness of understanding in conversation. It contains induction as an approximation: it explains something that looks like induction.

Similar to biological evolution, which does not have a goal, the evolution of a function pool does not have a goal either. Yet both create complex adaptations, which, in hindsight, appear goal-directed – one over millions of years in the form of biological adaptations, the other in a single lifetime in the form of ideas.

Likewise, the non-random differential reproduction of functions contains the mistaken Lamarckian theory of the origin of knowledge through use and disuse as an approximation: those ideas that are better replicators end up dominating the population. What seems like induction through use is actually the greater differential reproduction of some functions over others.

The theory explains more things than that. It explains the deep connection between people's explanatory universality and computational universality: since a mind can evolve any function, and since functions are explanations, it can therefore evolve any explanation. The theory explains that problems are harder/easier to solve in terms of the number of successive variations required toward a solution, the competition between functions, and differential reproduction of functions. In other words, it explains the difficulty of a problem in terms of the distance in a function's tree of ancestors from the point the problem was detected to the point the function was detected as a solution. It shows why biological evolution and meme evolution, though likewise constituting evolution and error correction, are not conscious: because there is no meta-algorithm, no single and coherent "center for error correction" accompanying either. It explains how phylogenetically[22] acquired, innate dispositions interact with the mind at runtime, and it explains how ontogeny[23] alone cannot result in new knowledge without an evolutionary algorithm operating at runtime. It explains why animals lack this algorithm, and why people have it. It explains *understanding something* as an accidental mutation of a function, which mutation leads to the function's representation of (i.e., correspondence to) an actual process, for example, a physical one in the case of a scientific function. It shows how functions, including memes, lead to behavior both in creative and non-creative beings: by being invoked by the meta-algorithm.

Lastly, we can recover mistaken explanations of neuroplasticity in terms of the brain's hardware through the reconstruction of missing functions from copies and variations of the original. Consider, for example, the case of a man who lost between half and three quarters of his brain due to fluids building up in his skull (NewScientist, 2007). Lost brain regions included those allegedly responsible for "motion [control], sensibility, language, vision, audition, and emotional and cognitive functions." Yet, aside from "mild weakness in his left leg," the patient functioned almost entirely, even though the fluid had been slowly eating away at his brain for decades, leaving only a thin sheet of brain matter on the outside.

Rather than dismiss the mistaken notion of special-purpose brain regions in favor of the entire brain being general purpose, researchers concluded that the remaining brain region, though somehow presumably still special-purpose, must have taken over the missing functionality! As I discussed in chapter 4, such customizability is evidence of a *universal* system. What happened is that those functions stored in memory that was slowly being eaten away by fluids were instead continually reconstructed by creativity and stored elsewhere in memory that was still available. Creativity was able to reconstruct them from many copies and variants that were still around in the function pool. "Neuroplasticity" is a misleading term because it attempts to explain a property of creativity, namely the ability to recreate originals of ideas from their variants, in terms of hardware (neurons), even though it is a software matter.

This misplaced focus on hardware is widespread, including the misconception that in the brain, hardware and software are identical (see also chapter 7). In "The Selfish Gene," Richard Dawkins first introduces his idea of cultural units of replication: memes. As I have argued, a meme is a function that has "jumped" from one mind to another. He writes: "If memes in brains are analogous to genes they must be self-replicating brain structures, actual patterns of neuronal wiring-up that reconstitute themselves in one brain after another." But that is not so. Brain structures do not replicate – they do not have a way of getting from one brain into another. That would require surgery. And of course, the same meme can be stored in different brain regions in different people, just like the same program can be stored in different parts of a hard drive on different computers.

Further, the same meme need not necessarily result in the same wiring in two different brains. The meme of say, addition, may have different implementations in several different brains, because different people may explain addition differently (though they are all still *qualitatively* the same if they are good explanations). Their implementations further depend on their background knowledge, what they thought addition to mean, et cetera.

But even in the event that a meme replicates without any difference in implementation, explaining memes in terms of the underlying hardware is misleading. Memes are *software* – functions that spread from mind to mind. We do not explain them in terms of brain regions or neuronal wirings because that is the wrong level of emergence. It is reductionism again. Indeed, *all* replicators – including those behind genes – are software, and therefore functions, which causally affect the physical world. A replicator of any kind no doubt needs to be *instantiated* in a particular physical substrate – genes in DNA molecules, ideas in brains – but what that substrate looks like and what it is made of matters only to the degree that it can encode an instance of the replicator and help it spread. If replicators were not abstract, they could not copy themselves from one physical medium to another, that is, instantiate themselves several times in separate physical forms, because, as I have argued, that would entail cloning matter. Therefore, I believe it would be best if Dawkins abandoned reductionism: if reductionism were true, then the theory of evolution would be false. The two theories are incompatible because reductionism denies the possibility of the downward causation of selection pressures on individual organisms (Popper 1977).

While this theory of the mind seems to have good explanatory power, it is incomplete, or else we could build AGI tomorrow. It raises a few questions. I suggest those interested in building AGI consider them.

One such question is: how does the meta-algorithm work, in detail? The evolving function pool does not seem very difficult to implement; it is similar to evolutionary algorithms that have been built before. However, present-day evolutionary algorithms do not have a meta-algorithm that adopts newly conjectured criteria for criticism and submits the function pool to them, thereby introducing new problems and "stirring up" the function pool again. So, any solutions that had been converged upon are now also stirred up, and new solutions can be found, whereas present-day evolutionary algorithms do not yet have this ability to "stir up" the function pool and, therefore, often converge without producing any more results (see chapter 7 and also Deutsch 2012a, 161).

The difficult bit seems to be the interplay between the meta-algorithm and the function pool, and the policy according to which the meta-algorithm invokes and criticizes functions from the function pool. It is not clear yet how the meta-algorithm evaluates and adopts functions from the function pool as criteria for criticism and pruning. But it seems to. Otherwise, the meta-algorithm would have to itself be subject to change according to some even higher-level meta-algorithm, which would lead to an infinite regress. So whatever the meta-algorithm does, though static, must have enough reach to allow for the regulating interplay with the function pool, which interplay is itself subject to change.

Harder problems have been solved. A possible solution is that when a problem is found, the meta-algorithm's criterion is temporarily updated with a criterion for what would constitute a solution. Alternatively, the criterion is updated when a mutated function claims there is a problem with it. These are two ways in which the meta-algorithm's criterion for error correction is itself subject to change. And we need to figure out the pruning mechanism, too.

Perhaps one way to reverse-engineer the meta-algorithm is to think about what it is that people *cannot help but do*. That is because people, like any computer program, simply follow instructions.

While we can explain that the mind operates on the body by invoking conjectured functions that, in turn, invoke the body's interfaces, we do not yet know how the body acts on the mind. For example, how does the body signal to the mind that a wound hurts? (I am looking for explanations on the same level of abstraction as creativity, not in terms of hardware such as nerves. Nor am I looking for an explanation in terms of a quale, but instead for the underlying process that gives rise to that quale.)

Since empiricism is false, we know that AGI can work without any inputs, so we can build the function that is AGI without any arguments. It is helpful to exclude inputs in the implementation because it helps generally in software engineering to reduce a problem to the necessary parts first.

I guess that an AGI can start with a single instance of a self-replicating function in its function pool, as long as mutations can, in principle, lead to any other possible function (within memory-capacity constraints). Thus, over time, though every person starts with roughly the same meta-algorithm, the same policy, and the same function pool at birth, they develop into different people because their function pools evolve differently. That is because evolution generally is unpredictable. After all, mutations are not deterministic, which is also why the *growth of knowledge* is unpredictable (Popper 1983, 298).

In other words, it should be fine to start with the same code for each new AGI: they will automatically develop into different people over time as they run. The details of the self-replicating function may not matter so much because it mutates away from its initial state relatively quickly.

Lastly, we know that AGI is an algorithm that does not terminate, so it either contains a loop or calls itself recursively. (When people in human bodies die, or "terminate," it is not because of some programmatic termination condition, but because the program is interrupted by bodily decay leading to death – in other words, the computer on which a person runs breaks.)

The genetic mutation that introduced functional replication in the mind was the spark required to light the fire that is creativity. It was a "jump to universality." (Deutsch 2012a, 146) The more a creative mind learns about the universe, the more it resembles it. Every mind is a little copy of the universe. Some copies are more accurate than others. Over time, as the mind understands the world better, its contents resemble a bigger and more accurate slice of the universe.

We thus see that creativity is a phenomenon that emerges from the underlying interplay of a meta-algorithm and a pool of self-replicating functions. The meta-algorithm exerts changing selection pressures onto the function pool, which is how different ideas evolve and how the mind comes to understand the world.

I suggest that AGI can be built by recreating this interplay. I further suggest that if this theory is found to be false, a theory *like it* will be needed to build AGI (as opposed to current narrow AI research; see chapter 7). A theory would be needed that explains how minds create knowledge (without ever running that theory as a program (Deutsch 2012b)).

Terminology

Intelligence – Creativity

Side effect – A change in the behavior or state of the outside world caused by a function

Phenotype – A function's side effects

Explicandum – A thing to be explained

Universal law – Function

Cause – Input/function parameter

Effect – Output/return value

Background knowledge – Knowledge currently deemed uncontroversial (Popper 1983, 71)

Determinism – "All events are determined completely by previous causes: one thing follows another like clockwork."

Indeterminism – "Events do not have prior causes, at least not deterministic ones."

Reductionism – The misconception that we need to explain things in terms of lower-level constituents (Deutsch 2012a, 123), and that causality can only move from the small to the large, not the other way round

Dualism (better: *pluralism*) – The true theory that both material and abstract things are real and interact

Panpsychism – The mistaken doctrine that every object is conscious

Function pool – The set of functions available to an organism

Meta-algorithm – A function overseeing the scheduling, execution, and interruption of functions from the function pool

Policy – A set of rules inspected by the meta-algorithm to determine which functions to run

Allele – A slight variant of a replicator

Biological evolution – Alternating variation and selection of replicating genes in a gene pool

Functional evolution – Alternating variation and selection of replicating functions in a mind

Proto-creative – A mind is proto-creative if it contains a functional replicator but cannot creatively detect errors

Fully creative – A mind is fully creative if it contains a functional replicator and can creatively detect errors

Thought – When the meta-algorithm invokes a function to find and correct errors

Memory – A phenomenon emerging from the non-random differential reproduction of function replicators in a mind

Summary

Intelligence and creativity are synonymous.

Creativity can create all kinds of knowledge, including but not limited to moral, emotional, and aesthetic knowledge, because it produces functions, and functions are computationally universal.

Creativity can feel inductive when a problem is solved fast, but that is always misleading.

Children are masters of creativity: they joyfully and seemingly effortlessly create vast amounts of sophisticated knowledge, such as how to use their bodies, how to interpret sense data, and how to speak.

Knowledge of how to use the senses is not encoded genetically, nor can it come from the senses; that leaves creativity as the only possible source of human knowledge.

All mental activities of people can be divided into the inspiration phase and the perspiration phase (Edison in Deutsch 2012a, 36). Creativity lives in the inspiration phase. There is no reliable, mechanistic way to create knowledge. Luck is involved.

To explain a problem is to create an explanation from which one can deduce the problem.

We have no direct access to reality; all observation is "theory-laden." (Popper) Neither the detection of a problem nor finding a solution is ever goal-directed.

Popperian epistemology misses the notion of a replicator, without which it runs into infinite regresses, deviates from neo-Darwinian evolution unnecessarily, and cannot explain the origin of conjectures. The introduction of functional replicators solves these problems.

Reductionism leads to bad explanations generally and bad explanations of free will and consciousness in particular. Consciousness is an epi-creative phenomenon related to the detection and correction of errors and disappointed expectations (Popper 1983, 344). Animals are not conscious or intelligent, nor do they have free will, because they are not creative. Animal welfare is not an issue. The presence of sophisticated knowledge is not evidence of intelligence because it can originate in biological evolution. Knowledge that is not encoded in the animal's genes and that the animal, therefore, must have created itself is.

Creativity evolved from non-creative ancestors through a single genetic mutation introducing a replicating function in the mind. Function replication, variation, and selection introduce non-genetically-determined knowledge in minds at runtime, including new criteria for error detection. A mind searches the space of possible ideas through imperfectly replicating functions: the explanatory universality of people is based on the computational universality of functions. That is how a single one of our ancestors became conscious, and it is why our species exists.

This theory solves outstanding problems with Popperian epistemology by explaining conjectures and criticism as adaptations that are themselves the result of functional evolution, and it explains additional phenomena such as memory. Further, it provides a research program for building AGI. An outstanding problem in this research program is: how does the meta-algorithm work?

Now that we have a research program for AGI, we can evaluate present-day intelligence research. But first, I want to introduce a profound truth about reason and software engineering.

6
All People Are Software Engineers

...and all people are philosophers (Popper 2018a). In the introduction to this book, I pointed out that software engineers routinely contribute to epistemology by finding rules governing the structure of software (Temple 2010). In chapter 3, I expanded on this view: a good program implementation, just like a good explanation, is hard to vary "while still accounting for what it purports to account for." (Deutsch 2012a, 31) This connection is no accident. It arises because program implementations, which I also refer to as functions, *are* explanations.

A programmer's mission is to solve problems by creating new knowledge, making it as explicit as possible, criticizing its structure repeatedly, writing it down, and then running it repeatedly to test it. I have also expanded on the properties explanations and functions share: a function's parameters are an explanation's initial conditions or its experimental setup for testing; a function's body is the explanation's structure; a function's return values are the explanation's claims, and we run functions to create claims and detect problems: conflicts between two or more functions. Since functions and explanations share all of these properties, explaining explanations in terms of functions is itself hard to vary. Thus, I concluded, this opens up a new way of explaining explanations: they are computations that are adapted to solving problems.

However, the fact that functions are explanations is just a corollary of a much bigger principle I call the *equivalence principle* (of epistemology, not to be confused with the equivalence principle of general relativity). It states that *software engineering and reason generally are equivalent*. It is why so many software engineers say that the craft helped them *learn how to think*. Reason is the quest for good explanations, scientific or philosophical, and science is that part of philosophy whose good explanations make predictions that are testable by observation.

Software engineering represents all of reason: all programs are explanations of how to solve problems, and all explanations are programs. Some programs are good explanations; some are bad ones. Implementations that can be tested by observation are scientific: they are tested by checking if their return values match those of implementations interpreting sense data. The question "how do we know?" is the same as "what is the origin of computer programs?". The answer is creativity. (Though a more productive question than "how do we know?" is, "how can we detect errors in our programs?".) And the question of how creativity works is the same as the question of how to build AGI.

As I have said, Xenophanes, Democritus, and other Greek philosophers were not aware they were working on AGI, but that is what they were doing. (Of course, had they made a breakthrough, they would have lacked the concept of universality, the theory of computation, and physical, general-purpose computers to *program* an AGI. But they would have known how to do so in principle, without running it as a program.) Similarly, I am not aware of Popper ever stating his work in terms of AGI. But in his time, computers were available to build one.

We already explored many of the corollaries of the equivalence principle in chapter 3. Generally speaking, we can divide them up as in the following table:

The Equivalence Principle of Epistemology

Reason	Software Engineering
Explanation/theory/knowledge	Function implementation/ program
Good/bad explanation	Function implementation whose structure is hard/easy to vary "while still accounting for what it purports to account for." (Deutsch 2012a, 31)
Data	Return values

Reason	Software Engineering
Philosophy	Solving problems by writing function implementations
Science	Solving problems by writing empirically testable function implementations
Philosopher	Software engineer
Scientist	Software engineer who writes good, empirically testable function implementations
"How do we know?"	"What is the origin of computer programs?"
Creativity/personhood	AGI
To understand/explain/create knowledge/solve a problem	To write a function implementation
To know	To have a function implementation
Problem (conflict between explanations)	Conflict between functions
Criticism	Finding conflicts between functions
Darwinian evolution	The creation of functions through alternating variation and selection
Neo-Darwinism	Functional Darwinism: a theory of replicating functions and non-random differential reproduction of functions

Reason	Software Engineering
Induction	The mistaken idea that functions can be created by extrapolating from return values
Lamarckism	The mistaken idea that functions can be engineered through use and disuse of other functions
Meme (Dawkins 2016, chapter 11)	A function that replicates across holders
Memeplex/"meme-complexes" (Dawkins 2016, chapter 11)	A group of functions that help cause each other's replication across holders

Written software code is a concrete, physical manifestation of knowledge. This connection goes deeper than meets the eye. At first glance, software code merely *contains* knowledge rather than *being* a concrete manifestation of it, just like books contain knowledge in the form of letters and punctuation marks, which readers can decode and turn into mental concepts using their knowledge of how to read. As I have stated before, I disagree here with Popper who claimed that books literally *contain* knowledge (1983, 107). Books contain knowledge only indirectly: a mind needs to decode that knowledge through guesses, which, if they are good guesses, approximately instantiate the knowledge the book's author used to write the book. Software stored on hardware, on the other hand, *is* knowledge and requires no mind to be recreated first.

David Deutsch has settled on a simple test of understanding: "If you can't program it, you haven't understood it." (2012a, 154) In other words, if you understand something, and you know how to write computer programs, you can replicate what you have understood in the form of a computer program.

This test is an excellent yardstick to gauge someone's understanding of something, and, given the conjecture that function implementations are explanations, I would like to reformulate it in a stronger version: To understand something *is* to program it, if only mentally. This equivalence gives us a *criterion of understanding*. It does not mean that people who are not software engineers cannot understand anything. Instead, it means that to understand anything is to mentally implement a function that replicates that thing in our minds. It is also why I said that all problems are soluble via software: even solutions that cause physical transformations are first created as software, and then executed as software, in a mind. The equivalence principle is the broader concept from which Deutsch's test of understanding follows as a corollary.

I am aware of three criticisms of the equivalence principle of epistemology.

The first criticism says that most scientists do not primarily write software. They may do so on occasion when it helps them run or evaluate an experiment, but it is only a tool they use. Philosophers use software even less, if ever; they spend most of their time thinking. But that is misleading. Philosophers and scientists – all people, in fact – encounter problems they want to solve, which they always do by creating knowledge in their minds. Those minds run on computers – their brains – and conjecture solutions to those problems in the form of computer programs. So the fact that all programmers are philosophers is just a special case of the general truth that all people are philosophers and that all problems are solved via software: knowledge. Programmers do this most obviously and explicitly. Coming up with a hard-to-vary explanation is always a software-engineering effort, either in a mind or in genes. Whether a non-brain computer is involved and whether the conjectured software is eventually written down in a text editor on that computer is another matter.

The second criticism claims the equivalence principle is reductionist: how can higher-level concepts and explanations possibly be encoded in low-level entities such as functions? The answer is that, following from the computational universality of Lambda

Calculus, functions can represent concepts on any level of emergence. They are not necessarily low-level only.

The third criticism states that the equivalence principle's functional approach only represents one mode of explanation when, in reality, there are many different modes. One mode of explanation in physics that I mentioned in the previous chapter is that of initial conditions and laws of motion. In contrast, another mode is that of laws of physics and the physical transformations they allow and forbid (Deutsch 2013). Supposedly, the mode of explanation that functions represent resembles that of initial conditions and laws of motion: a function's parameters are said to represent initial conditions and the function's body a particular law of motion.

While we can indeed think of a function's parameters as initial conditions, and its body as a mapping of those parameters onto a result, interpreting its body as a law of motion is a mistake. Instead, the particular mode of explanation a function incorporates is reflected in its body's structure and the way people write software. This structure can be influenced by conventions that can change over time, much like they do in the sciences. If a function takes as parameters an object `a`, its position `b`, and a vector `c`, then it can indeed compute `a`'s next position. But if instead, a function takes as parameters the laws of physics as we know them `a`, and an object `b`, it can compute which physical transformations of `b` the laws `a` allow.

Alternatively, consider a third mode of explanation: the evolutionary one, in which functional replicators mutate and compete to spread through a population of replicators. This mode is the one needed to build AGI, and we know from computational universality that we could build it using the functional approach. Of course, in all three examples, those functions need to contain the requisite knowledge in their bodies to perform the desired computations. But the point is that these examples represent all three modes of explanation.

Further, it follows from the universality of computation that *all possible* modes of explanation can be expressed through functions, and, not surprisingly, that it is possible to write higher-level functions *explaining* different modes of explanation. I guess that any successful refutation of the equivalence principle will also need to be a successful refutation of the universality of computation. One could refute it by, say, showing that the evolutionary mode of explanation, and with it, creativity, cannot be programmed in Lambda Calculus.

Terminology

Equivalence principle (of epistemology) – Software engineering and reason are equivalent

Weak criterion of understanding – "If you can't program it, you haven't understood it." (Deutsch 2012a, 154)

Strong criterion of understanding – To understand something *is* to program it, if only mentally. "Have you programmed it, at least mentally?"

Summary

Understanding something is the same as programming it, if only mentally. Software engineering and reason are equivalent. All people are both philosophers and software engineers because they all write software, at least mentally, and that software consists of explanations.

Now let us evaluate present-day intelligence research.

7
Current Intelligence Research

Intelligence is a kind of knowledge: it is the creative algorithm. Creativity is knowledge that is adapted to the general purpose of solving problems of any kind. I use the terms creativity and intelligence interchangeably because they are the same thing. People have hoped to recreate intelligence on a computer for many decades, but they have made no progress (Deutsch 2012b). It is not for lack of trying, but due to severe misconceptions that have led researchers down blind alleys. I will discuss how these misconceptions influence intelligence research and explain why they cannot work, why they actively inhibit progress, and why we need to abandon them in favor of Popperian research programs.

Popperian epistemology is our best explanation of intelligence, which means that any effort not devoted to translating creativity into a computer program is futile (Deutsch 2012b). This insight alone already allows us to reject most of present-day intelligence research. As I will show, Popperian epistemology conflicts with all other purported explanations of intelligence and research programs.

Rejecting other explanations in bulk is usually looked down upon as black and white thinking. Relativists want to remind us that while we may have a point here and there, it is essential to consider rival explanations as well and *compromise*. It is indeed essential to consider rival explanations, but not after those explanations have been refuted. And it is never a good idea to compromise: since good explanations are discrete, merely picking something halfway between two good explanations does not lead to another good explanation (Deutsch 2012a, 342). It usually leads to garbage.

So it is not clear what relativists expect us to do. Consider worse explanations just for the heck of it? Relativists have no interest in discovering how creativity (or anything) works; they are only interested in signaling how tolerant they are.

Relativism is an *anti-rational meme* (Deutsch 2012a, 388-390): it actively undermines criticism. Part of the creative process is to come up with many rival explanations and eliminate them one by one until the best one survives, and then to act on it without hesitation. Therefore, taking sides with relativists against going with the best available explanation – Popperian epistemology – is to cause AGI research never even to get started.

Relativist behavior need not be overt: it can seem as innocuous as thinking to oneself "huh, how interesting" after reading a refutation of an idea, and then acting according to that idea anyway. I have always seen the relativist attitude as a betrayal of reason, and an opposition to progress and the alleviation of suffering, for we have a moral obligation not to spread ideas which we should know to be bad explanations.

In order to show where narrow AI research is going wrong, we need to apply severe criticism. There is no way around it. As Popper argued, "[…] we want our criticism to be *severe*. In order that the criticism should be severe we must use the full apparatus; we must use all the guns we have. Every shot is important. It doesn't matter if we are over-critical: if we are, we shall be answered by counter-criticism." (Popper 1983, 305)

The growth of knowledge is unpredictable (Popper 1983, 298), and one of the existing fields may surprise us with a solution. But I would not hold my breath. It is important to understand others' mistakes so we do not repeat them or head down blind alleys, too. That said, we will enthusiastically refute many different approaches.

Narrow AI Research

As I stated in the introduction, artificial intelligence research is different from artificial *general* intelligence research in that it is *narrow*. That does not mean it does not have its merits; it has been getting much attention for a reason. It achieves results that were not possible before. However, it is targeted at concrete applications, such as processing language, competing at chess and Go, and autonomously driving a car. Therefore, its reach is inherently limited, whereas AGI has infinite reach: it can solve any soluble problem.

"AI" as a term used to mean human intelligence. Then people bastardized the term to refer to programs that have passed a level of sophistication, or worse yet, to programs that perform tasks that traditionally only humans could perform. The latter definition is useless because *all* technology is an effort to automate things humans do so that they can focus their creativity on new problems. But AGI is a radically different program because it will be the first and only technology, not just computer program, ever to solve problems rather than automating solutions. Narrow AI lives in the perspiration phase, whereas AGI lives in the inspiration phase. So AI is not only narrow in reach, but has a different character: it is not about creating knowledge.

Consider, for example, the field of *machine "learning."* One of the simplest tasks in this field is that of *regression analysis*, or "curve fitting." For several labeled points in a coordinate system, regression analysis can predict the corresponding y-value for a new x-coordinate. For example, consider a plot of the dollar values of houses by square footage. (Other factors determine the value of a house, such as neighborhood, school districts, year of construction, et cetera, but the mechanism does not change as we introduce additional dimensions.) A curve-fitting algorithm works by finding the curve that is the closest to all given points, usually through a mechanism called the "least-squares method": it minimizes the sum of all the squared distances to each point. It then uses that curve to predict the y-value for a new x-value.

This mechanism works for linear data (we would expect the values of houses to stand in linear relation to their square footage) as well as non-linear data. With the generated line, you can now predict the price of a home given its square footage. This approach has reach: you can use it to write programs that can predict any value given arbitrary data.

However, note that it is *people* who pick data according to whether or not it affects the value (Temple, personal communication with author, 2019). For example, it just makes sense that square footage is a predictor of the price of a house. The algorithm will try to predict the price of a house given the previous owner's hair color if instructed to do so. It could not tell that this would not make sense. It is people who decide to configure the algorithm, and it is people who specify the optimization via the least-squares method.

Note also that this program can solve one class of problems only, namely regression problems. Further, it cannot detect new problems, let alone solve them. But detecting new problems of any kind and solving them is what we expect a creative algorithm to be able to do.

A common and more sophisticated task for a machine learning system is to recognize handwritten digits and transcribe them to computer digits. That is an example of a *classification algorithm*. How could this be achieved?

First, people label a large number of images containing handwritten digits and then ingest them into a program superficially resembling a brain-like *neural network* consisting of neurons arranged in separate layers. For each given image, the neural network returns the digit it recognizes in the image, which is compared with each label, allowing the program to fine-tune the network's weighted connections to increase accuracy through so-called feed-forward and back-propagation algorithms. This phase is called the training phase. These algorithms are powered by linear algebra (the neural network and its weighted connections are just a visual representation of matrices) and calculus: a so-called gradient-descent algorithm finds a configuration of the neural network that leads to minimum mismatch.

In other words, the gradient descent finds a local optimum. But optimization does not constitute learning. One core fallibilist tenet is that there can be no source of truth or closeness to truth (Popper 2002b, 3-39). Therefore, intelligence cannot possibly be an optimization problem, in which an optimization function is known and acts as a source of truth. There is no computable criterion for success in the creative algorithm. Any such alleged criterion is itself born out of conjecture and subject to error and criticism, and therefore nothing like a source of or criterion for truth.

Recognizing handwritten digits is a more sophisticated kind of curve fitting: both share the underlying mechanisms. Note that this program has some reach: it can, in principle, be trained to recognize anything; however, once it is trained, it can only be used to recognize what it is trained to recognize. Alas, the whole mechanism has an inductivist flavor: a solution to a problem (knowledge) is said to be somehow obtained through repeated observation. This repetition constitutes a regularity, which is said to create knowledge. But the creation of knowledge starts with problems which can arise from regularities (sunrise) as well as irregularities (midsummer night), or entirely novel situations (solar eclipses).

From an inductivist point of view, a neural network seems like the obvious solution. However, we know that induction cannot lead to knowledge creation, so the better explanation for the resulting knowledge of how to recognize handwritten digits is that it was part of the reach of knowledge humans created first and then translated into a program (see also Deutsch 2012a, 161-162). Indeed, before running the program, humans had to determine the number of layers and nodes in the network, the initial weights between nodes, and the correctly labeled training data. Calculus and linear algebra are also kinds of human knowledge with great reach that people have purposefully leaked into the program. The program discovers no new problems. It is not creative and, therefore, not intelligent. It is admittedly sophisticated but still merely the automation of a predefined task.

To an inductivist, it makes sense to think that babies learn by observing the world and taking in massive amounts of data like sponges. But you cannot observe without a theory of what to observe. Even the term "data" ("givens") is misleading because they are not given. Only from theory can we know "what to look for and how and why" (Deutsch 2012a, 18); that is, what "data" to collect. Data are return values of functions one must conjecture first. Observational data are those returned by theories governing the senses. But a human brain can conjecture and refute ideas even when in a vat (Deutsch 2012b); no input or output is required to be creative. Real learning happens by solving problems. No amount of ingested data can lead to creativity. As Popper says in "The Myth of the Framework" (1996, 7-8):

> The inductivist or Lamarckian approach operates with the idea of instruction from without, or from the environment. But the critical or Darwinian approach only allows instruction from within – from within the structure itself.
>
> In fact, I contend that there is no such thing as instruction from without the structure, or the passive reception of a flow of information which impresses itself on our sense organs. All observations are theory-impregnated. There is no pure, disinterested, theory-free observation.
>
> [...]
>
> We do not discover new facts or new effects by copying them, or by inferring them inductively from observation, or by any other method of instruction by the environment. We use, rather, the method of trial and the elimination of error. As Ernst Gombrich says, "making comes before matching": the active production of a new trial structure comes before its exposure to eliminating tests.

When Popper mentions "the method of trial and the elimination of error," he speaks of an evolutionary algorithm. It is the kind of algorithm I discussed in chapter 5, and I will review current approaches to this class of algorithm below.

There are so-called "unsupervised" learning algorithms that take unlabeled data. One use of such programs is to structure data into different clusters. For example, a vacation rental website may want to show listings by neighborhood. Clustering is another kind of classification algorithm. Since the data are not labeled, it may seem like less human knowledge is leaked into the program. But in reality, the labels are replaced by rules that have reach, such as a clustering algorithm called *k-means*. It is, again, people who provide these rules, and they contain all the knowledge required to divide the data into separate clusters.

Philosophers investigated the role of classification centuries before software engineers did. Strict empiricists advise us to interfere as little as possible with pure sense data in order to accumulate pure knowledge. If we interfere with perception through our prejudices and preconceptions, they say, it leads to error. Kant, on the other hand, denied that perceptions are ever pure, and claimed that our experience is the result of perceptions undergoing some "[automatic] processing," such as "systematic classification." (Popper 1983, 342) What is missing in all of these cases is an explanation of how the knowledge of how to classify was created in the first place. There is no mention of evolution. (We should forgive Kant since the theory of evolution was not discovered until after he died.) Additionally, classification has limited reach: classification algorithms can only solve classification problems, and not all problems are classification problems.

Another kind of "learning" algorithm is *reinforcement "learning."* It is used, for example, to "teach" computers how to play games. In December 2013, researchers from the London-based AI company DeepMind released a paper called "Playing Atari with Deep Reinforcement Learning." (Mnih et al. 2013) A little over a month later, in January of 2014, Google announced they had bought DeepMind for $500 million. DeepMind made the news in 2016 when its program "AlphaGo" beat the professional Go player and world champion Lee Sedol.[24] In 2017, they introduced "AlphaZero," a program with greater reach that can play a variety of games. It mastered the games of chess, Go, and shogi (a popular chess variant in Japan) and defeated all world-champion programs in these games after 24 hours of playing against itself. It also defeated its predecessor, "AlphaGo Zero."[25]

Reinforcement learning systems "teach" an artificial player how to behave in an environment based on how well it is performing. The player's goal is to achieve the maximum reward following a given policy (not to be confused with the meta-algorithm's policy from chapter 5). For example, the maximum reward in the game of chess is to win the game. The actions dictated by the policy can be chosen probabilistically – for example, according to Bayesian rules.

This functionality is inspired by Bayesian epistemology and results in having scores, which Bayesians prefer because they can compare them much more easily than explanations. Translated into reinforcement learning, it says that a rational player is more likely to choose actions that are rewarded highly.

While Bayesian statistics is a useful field, Bayesian epistemology is another form of inductivism and therefore mistaken. Additionally, reinforcement learning is influenced by behaviorist psychology and inspired by animal "learning." To think this is how humans learn is a behaviorist mistake. Even though mainstream psychology now rejects behaviorism, computer science holds on to it because behavioristic input-output models can be useful for many types of programs that are not AGI (Deutsch 2012b).

I discussed behaviorism in chapter 3; it is the mistaken doctrine of instrumentalism applied to psychology. It is also a widespread mistake in the raising of children: they do not learn through reward or punishment. They learn by solving problems.

For all these reasons, we must conclude that reinforcement learning does not constitute knowledge creation. Reinforcement is a way to get a system to do precisely what you want it to do. It is a mechanism to enforce compliance, not to create knowledge. Besides, a creative being chooses its rewards. In a reinforcement learning algorithm, the knowledge of what to consider reward and punishment, as well as the mechanisms by which to seek out reward and avoid punishment, are already present. In the case of an animal, they are given genetically; in the case of reinforcement learning on a computer, they are given by the programmer. In both cases, the requisite knowledge is the result of evolution happening before runtime, either through biological evolution or evolution in the programmer's mind (see also Deutsch 2012a, 161-162). No evolution is happening in the program. Therefore, attributing intelligence to narrow AI is the same mistake as attributing it to animals, no matter how sophisticated the knowledge in question may be (see also Temple 2019b):

> Biological evolution ——> Sophisticated knowledge encoded genetically in organisms such as animals, some of which manifests as behavior (part of the organism's phenotype)
> Evolution in people's minds ——> Sophisticated knowledge encoded digitally in programs, some of which manifests as behavior (part of the program's phenotype)

In neither case does evolution occur on the right side of the arrow. But only that would constitute intelligence. Thinking that the mere presence of knowledge is evidence of intelligence – reasoning of the sort "there can be no design without a designer," and that the designer in this case must be the animal or narrow AI itself – is an ungainly remnant of creationism.

In all of the above examples of machine learning – regression analysis, image recognition, and reinforcement learning – the shared misconception, *by definition*, is that learning somehow happens through experience and by drawing conclusions from data, with mechanisms for how to draw such conclusions implemented by the programmer. The focus is on data and patters rather than discovering problems and creating knowledge. The underlying implementation – for example, whether an algorithm leverages a neural network or something else – is incidental. Real learning is a never-ending cycle of problem detection and attempted solutions through trial and error correction. Only creativity can achieve this, and it happens with or without experience, that is, input data.

Inductivism has had a negative influence on computer science ever since its beginning. The following are excerpts from a lecture Alan Turing gave at the London Mathematical Society (1947). Interestingly, he mixes many prescient remarks[26] with bad misconceptions (a combination we will encounter again in one of his papers later on):

> It has been said that computing machines can only carry out the processes that they are instructed to do. This is certainly true in the sense that if they do something other than what they were instructed then they have just made some mistake. It it also true that the intention in constructing these machines in the first instance is to treat them as slaves, giving them only jobs which have been thought out in detail, jobs such that the user of the machine fully understands what in principle is going on all the time. Up till the present machines have only been used in this way. But is it necessary that they should always be used in such a manner?

Here, Turing suggests that the everyday use of technology need not be its only use. In chapter 2, I made a similar suggestion, saying that the way we program need not be the only way and that AGI may fall out of other modes of programming, in particular ones whose goal is error correction rather than automation. So far, so good. He continues:

> Let us suppose we have set up a machine with certain initial instruction tables, so constructed that these tables might on occasion, if good reason arose, modify those tables. One can imagine that after the machine had been operating for some time, the instructions would have altered out of all recognition, but nevertheless still be such that one would have to admit that the machine was still doing very worthwhile calculations. Possibly it might still be getting results of the type desired when the machine was first set up, but in a much more efficient manner. In such a case one would have to admit that the progress of the machine had not been foreseen when its original instructions were put in.

Here, Turing speaks of the reach of such instructions. He continues:

> It would be like a pupil who had learnt much from his master, but had added much more by his own work. When this happens I feel that one is obliged to regard the machine as showing intelligence.

Here he makes the first mistake: the ability to alter instructions to the point of being unrecognizable is not sufficient to make a program intelligent. While an AGI's function pool *would* move away from its initial state beyond recognition, not every program whose state changes beyond recognition is an AGI. He confuses the reach of the instructions with creativity. While a person can make modifications to his software, and does so frequently, this ability alone does not make him intelligent. Turing then goes on to make his second mistake:

> [...] What we want is a machine that can learn from experience.

It is an empiricist mistake. On the surface, machine learning has achieved what Turing asked: for example, a neural network's weights are changed (instructions are modified) during the training phase (experience) to recognize something in images. However, as I have argued, actual learning from experience in the sense of knowledge creation is impossible. Real learning always works by the method of trial and error correction, with or without inputs. I guess that the present-day idea that machine learning constitutes intelligence is a continuation of Turing's mistake above.

Machine learning is just one branch of narrow AI research. Other areas include that of personal assistants, which I will cover below. There are several other implementations of narrow AI, and I expect people to build many new ones for concrete applications in the future. Are these programs useful? Yes. Are they intelligent? No. They do not create knowledge but instead do what all other technologies do: they automate solutions that were created by people. Therefore, narrow AI is the opposite of AGI, which *creates* solutions.

Many companies implement narrow AI to solve problems. Tesla is such a company: at the time of writing (September 2019), their goal is to build self-driving cars. Others use machine learning to recognize faces, or to classify spam, or to recommend products. One could argue that they do not pretend to build intelligence. In that case, they should have no problem to stop calling it machine "learning" and artificial "intelligence." These terms are false and misleading, thereby damaging and distracting from genuine intelligence research. It is virtually impossible to speak of intelligence research without someone asking, "oh cool, you mean, like, machine learning?", when that is just the opposite.

Moreover, I guess that many companies very much rely on these terms for marketing in order to brand their products as more sophisticated than they are. I suggest using the term "smart" to replace the term "intelligent" to avoid confusion. A program is *smart* if it contains sophisticated knowledge. But that says nothing about the *origin* of that knowledge, no matter how sophisticated. It may well have originated from biological evolution (see animals, which are also smart but not intelligent) or from the evolution happening inside the programmer's mind. A program is *intelligent* only if it *creates* knowledge by itself. Companies correctly market sophisticated phones as "smartphones," not as "intelligent phones." Why do they not call this kind of technology generally "smart technology" rather than "AI"?

But it gets worse. Some companies claim to work on AGI but do the opposite. One prominent self-proclaimed AGI researcher is "OpenAI" in San Francisco. Their mission statement (2019) is as follows:

> OpenAI's mission is to ensure that artificial general intelligence (AGI) – by which we mean highly autonomous systems that outperform humans at most economically valuable work – benefits all of humanity. We will attempt to directly build safe and beneficial AGI, but will also consider our

> mission fulfilled if our work aids others to achieve this outcome.

Recall that AGIs are people. OpenAI's definition of AGI is wrong. People are not "highly autonomous" – they are *entirely* autonomous. It is a contradiction in terms to say that people can outperform people, but whatever system they are hoping to build, it cannot possibly outperform people. As I said in chapter 4, people, by definition, have reached universality in the domain of general-purpose problem solving. There is no greater reach than this; outperforming it is impossible. I will cover safety concerns in the next chapter.

What they mean by "AGI" is just another kind of narrow AI, because they want to build a program that does "economically valuable work" – in other words, more perspiration, not inspiration. Looking at the work they showcase on their website supports this view: among other things, they have created a musical composer, a text predictor, and a piece of hardware that looks like a human hand that can manipulate objects. You could put all of them together to make a system that can do all three things, but that does not get you closer to the infinitely big repertoire that universal explainers (Deutsch 2012a, 146) have. It also does not get you creativity.

They have a section on their website called "OpenAI gym," which is an environment to develop and run AI systems. It features examples of specific algorithms and so-called "toy problems" and thus entrenches the alleged values of specificity in the minds of young contributors to the field. It appears that OpenAI does not understand universality or epistemology. Both are required to build AGI. Microsoft invested 1 billion dollars in OpenAI. Would they not like to know where OpenAI is going wrong to know if their money is well spent?

I mentioned London-based DeepMind previously when I referred to their reinforcement learning programs "AlphaGo" and "AlphaZero." Their mission statement says (2019):

> We're on a scientific mission to push the boundaries of AI, developing programs that can learn

> to solve any complex problem without needing to be taught how.

Such a program sounds like a universal problem solver – AGI. But they do not make good on that quote. They write papers in particular domains, and they do not work on general-purpose problem solving, only on the automation and execution of particular tasks. If we are trying to build AGI, we must consider specificity our enemy. It is possible that making a breakthrough in a specific field can open that field up and make its repertoire explode to infinity. But when this is not the goal, people go out of their way to constrain the program only to perform the specific purpose for which they designed it. That is the kind of situation AGI researchers find themselves in today. If one of them tells leaders in the field in Silicon Valley that he has an idea for how to build an AGI, they will tell him to narrow the problem down and run it in the OpenAI gym. This problem is no small one: it actively discourages striving for infinite reach in the domain of problem solving.

The theory of evolution is a mode of explanation. As I have discussed, rather than explaining how something *will* happen given some initial conditions and laws of motion, the theory of evolution explains how something *can* happen. Specifically, it explains how complex adaptations can come into existence without explaining how any particular one of them *did*. (Popper 1983, 267) Narrow AI operates in the space of conventional modes of explanations: it is about creating an "intelligent" program that is guaranteed, or at least likely, to create solutions to problems given some initial conditions and a "problem-solving" algorithm. That cannot be creativity because creativity depends in part on *luck*.

The creative algorithm does not explain how any particular problem is solved. It explains how the mind solves problems *in principle*. Most scientifically minded people adopt evolution as the explanation for how complex adaptations arise in the biosphere. But at the same time, they insist that knowledge must be justified, or likely to be true, to be considered knowledge. Therefore, they fail to apply their standards of how to (allegedly) do science to their (correct) embracing of the theory of evolution. They should feel a need to resolve this inconsistency. Instead, their persistently induction-flavored research shows they are perfectly happy accommodating this cognitive dissonance in their minds.

A common misconception about science (shared by many scientists, unfortunately) is that science is about laboriously gathering many facts, and that, if gathered correctly, these facts allow one to "derive" a theory that is certain and always true (Davies 1967, 1). This misconception has been driven to new heights through the coupling of machine learning with so-called "big data," which are enormous datasets gathered by companies so they can be scanned for patterns and regularities. While in many cases having more data can help correct errors, it is a mistake to think that "learning" algorithms "learn" more the more data they are given. That is inductive thinking again.

As David Deutsch says: "At the present state of the field, a useful rule of thumb is: if it can already be programmed, it has nothing to do with intelligence in Turing's sense." (2012a, 154) Current research makes other valuable contributions but does not contribute to an understanding of intelligence. The very fact that there are so many different narrow AI algorithms for different tasks tells us that none of them is creative in the sense of being a universal problem solver. For, if one of them were, it would immediately replace all others.

Recall the universal printing press: writing ever more programs from scratch is just like carving out more printing plates when, in reality, we need to build something like a universal printing press of software. That is the reason an accumulation of narrow-AI functionalities can never lead to AGI. You could build a software system that can recognize faces, predict text, drive a car, and play chess masterfully. But no matter the size of such an accumulation, it would always remain finite and, therefore, infinitesimal compared to AGI's unlimited repertoire. Such an accumulation would also have a wholly different character from an AGI in that it would not be *creative*: all its tasks take place in the perspiration phase, and none can create new solutions to new problems. Software engineering today concerns itself almost entirely with special-purpose problems, and even solutions that have great reach are limited compared to the abilities of AGI.

Each present-day software-engineering problem is like a printing plate before movable type. Some printing plates may be particularly interesting, but they are just printing plates nonetheless. Movable type gets you all of those printing plates for free: an AGI is automatically a self-driving car, a chess player, et cetera (if it chooses to be). In chapter 4, I discussed that general-purpose creativity is an algorithm with infinite reach: it can solve any soluble problem. It is universal in the domain of general-purpose problem solving. People can process (and, more importantly, learn and understand) language, play chess and Go, and drive cars, and do infinitely many other things, all using creativity. It is the only algorithm needed. Therefore, narrow-AI research will achieve only the opposite of AGI.

Neuroscience

Cleveland, to Joe:

> I'll just read to you from "Jet." Look, this whole page is all the Wayans brothers sitting in different positions. How you suppose they come up with all them positions? You could give me a million years. I just ain't wired that way.
>
> – *Family Guy, episode "Brokeback Swanson," 2015*

Peter, Lois, and Brian:

> Peter, to Lois: "[…] we'll be back with a turkey faster than you can say, 'Jack Robinson.'"
> Lois (slurred): "Jack Rober… Jack Row…"
> Peter, to Brian: "She had a stroke that killed the part of her brain that says 'Jack Robinson.' We got plenty of time."
>
> – *Family Guy, episode "Turkey Guys," 2014*

Neuroscience is the study of the nervous system, which includes the brain. Since the brain is host to intelligence, it seems logical that studying the brain could lead to an *understanding* of intelligence. In turn, neuroscience is said to potentially help us implement AGI either in the form of hardware by building a brain from scratch, or in the form of software we could then run on a computer. Neuroscientific intelligence research has an advantage over narrow-AI research: it is genuinely concerned with how people learn rather than how to automate sophisticated tasks. Nonetheless, it will fail. That is because neuroscience focuses on the study of the brain, which is hardware. But creativity is not a piece or property of hardware. It is a piece of software.

I do not doubt that if we could clone a brain atom by atom, we would thereby clone the intelligence on that brain. In other words, I do not deny that creativity lives somewhere in that brain. But it would not help us understand how creativity *works*. Also, neuroscience sees the brain as a machine that processes stimuli from the outside world, which processing results in some output or action (this is what inspired artificial neural networks). This image of the brain makes neuroscience inherently vulnerable to empiricist and behaviorist mistakes. It is also a mistake to think that in order to build intelligence, one needs to model the brain's hardware by replicating it through something like a neural network.

We know that the brain is a general-purpose computer because we are universal explainers. That requires a repertoire only a general-purpose computer can support. Universal systems are not designed for specific tasks; they are instead configured by software. The brain is hardware, and the creative algorithm is a particular piece of software running on that hardware along with other software.

Software is an abstraction. It is independent of its physical substrate. It cannot run without *some* hardware on which to run, but it does not discriminate between different forms of hardware as long as they are general-purpose computers. This substrate independence implies that creativity can run not only on brains but also on other computers: present-day desktop computers, laptops, smartphones, et cetera. Since a running creative program is a person (Deutsch 2012a, 414), people could, in principle, inhabit all of these devices – if we knew how to write the creative program. That means the program that is you can run on your brain, but also on a desktop computer.

It is difficult to imagine two more different pieces of hardware than a brain and a desktop computer. One is a wet, organic piece of meat, the other a dry, inorganic piece of metal and silicon. Thinking that there is something special about the brain in terms of hardware, for example, the neuronal structure or the so-called "massive parallelism," that supposedly will tell us how creativity works, violates the universality of computation (Deutsch 2012b).

Due to universality, there cannot be particular brain regions responsible for all tasks humans can perform. Somewhere in the brain, there is a universal computer, which includes memory and at least one processor, like any other computer. So what? Whether this computer's parts are spread out across multiple brain regions or not does not matter.

In any case, not every distinct task could be mapped onto a specific brain region, even in principle. It is said, for example, that the frontal lobe is home to motor control, concentration, planning, problem solving, speech, and smell, and that the temporal lobe is home to facial recognition. It is possible that the brain still contains special-purpose machines inherited from ancestors whose brains were not yet general-purpose computers, and that such special-purpose machines may still be used for, say, motion. But there cannot be a separate region in the brain for every possible task humans can perform because there are infinitely many such tasks. What region in the human brain is responsible for making phones? Which region handles writing books? And which one allows you to make tomorrow's discovery, for which there could not yet be a dedicated brain region? Assigning functions to human brain regions is to deny creativity.

Sometimes, people say that the brain is an analog computer[27]; that computers are not made of neurons, or that brains are perception and prediction machines, whereas computers are "calculation machines." These statements are all violations of computational universality and are born out of a reductionist prejudice – with a dash of empiricism here and there. For some reason, many are invested in the idea that the brain is a *special kind* of computer. They will readily admit that it is *some* kind of computer, but surely it could not be as simple as a programmable, digital computer!

This belief is a similar kind of superstition as the belief that life's physical constituents are special, a widely held belief until the discovery that our body's molecules are made of the same stuff of which inanimate matter is made. People back then should have known that all things must be made of *universal* components, and therefore should have rejected this superstition outright *before* that discovery.

Today, this strand of idea lives on in the superstition that there is something special about the brain's molecules, structure, neurons, or whatever it may be. Large parts of academia have been infected by it: prestigious institutions and people with impressive degrees subscribe to it. As the case of the "specialness of the material of living things" shows, these superstitions only go away *after* they have been refuted empirically (though sometimes not even then); people do not drop them for being bad explanations alone (which they should). If we knew how to program consciousness and ran it on a computer made of chewing gum and vacuum tubes, those interested would then study the "special properties" of chewing gum in order to understand consciousness!

Studying parts of the brain or the brain as a whole will help us understand its hardware, which may help us diagnose and cure diseases, and is interesting in its own right. But it will not uncover its *software* for the same reason studying a computer's metal and silicon does not tell you anything about how the software on that computer works (Deutsch 2018a). For example: If you want to understand how a web browser works – which is a piece of software – you will learn, roughly, that it works by making requests to servers over the internet and parsing HTML to display the webpages it receives in response to the user. That is a rough but good explanation, and it is devoid of anything to do with the underlying hardware, meaning it is an explanation on a higher level of emergence. It is an explanation that does not require analysis into lower-level constituents. Explaining software in terms of processor instructions and flipping bits in memory is reductionist and makes for bad explanations.

Here is another way to look at it. Suppose one day we know how to write the creative algorithm. Somebody presents you with two computers. Their hardware is identical: they both have the same amounts of memory and processing power, the same architecture, et cetera. You are told that one of them has the creative algorithm installed, but you do not know which one. You are asked to find out only by comparing the two computers' hardware: but they are identical!

Further, the instructions for the creative algorithm might be stored only in our genes, not in our brains, so the brain is possibly the wrong place to look for them. Or, if it is part gene and part meme (Deutsch 2012a, 414-415), some of it is stored in genes, and some in the brain. In that case, exclusively studying the brain might reveal only part of the picture.

Many corollaries of universality refute common misconceptions about brains and minds. They are outside of this book's scope, so I will only mention a couple of them briefly. One such misconception is the idea that a person's brain fully develops only around age 25 (or any other arbitrary age). But even toddlers are universal explainers, so their brains must already be universal computers. Nobody is too young to do anything. Whatever is missing until the age of 25 can only be knowledge/software. The same logic refutes the notion of hardwired *talents*: when a skill is present, it is due to knowledge being present. When a skill is absent, the requisite knowledge is absent. If they persist, people can create whatever knowledge they require, or else they would not be universal explainers.

Several myths surrounded children's abilities when I was a child. Children of engineers were automatically expected to be good at physics, which is Lamarckism. Children who were good at math were said to be bad at languages and sports, and those who were good at languages were said to be bad at math. Such claims are ridiculous, but I suspect many carried this notion into adulthood and assumed they would not be good at specific skills. So they never tried acquiring them, triggering a self-fulfilling prophecy. Claiming that creating any particular knowledge is not possible because "I'm just not wired that way" is another way of denying creativity. Furthermore, to say that some children are born talented at, say, math, cannot be right because there was no selection pressure for our ancestors to evolve knowledge about math biologically. That knowledge is passed on memetically, not genetically.

The Palm founder Jeff Hawkins is one of the more famous proponents of brain study in AGI research. He and his research company Numenta correctly guess that in order to build machine intelligence, one has to understand how human intelligence works first. He also has a real interest in building AGI. He is ahead of other researchers in both respects. Some of the most substantial criticism he has received so far is that his theories generally lack a body of empirical evidence supporting his models, but that is a justificationist mistake on the part of his detractors.

Unfortunately, he does make two compounding mistakes. His first mistake is thinking he can understand intelligence by studying the brain – he conjectures that intelligence is located somewhere in the neocortex. This conjecture amounts to denying that software is separate from hardware, a myth widespread in brain studies. His second mistake is his adherence to empiricism. In his book "On Intelligence," (Hawkins and Blakeslee, 2004) he presents the theory of "hierarchical temporal memory" (HTM) that powers pattern recognition. It is a kind of neural network and is, therefore, a continuation of the mistake of thinking that learning happens through experience discussed above. More recently, Numenta put forward a new theory of how the brain works: the "Thousand Brains Theory of Intelligence" (Hawkins and Maver, 2018):

> The Thousand Brains Theory of Intelligence proposes that rather than learning one model of an object (or concept), the brain builds many models of each object. Each model is built using different inputs, whether from slightly different parts of the sensor (such as different fingers on your hand) or from different sensors altogether

> (eyes vs. skin). The models vote together to reach a consensus on what they are sensing, and the consensus vote is what we perceive. It's as if your brain is actually thousands of brains working simultaneously.

This passage is empiricism ad nauseam. In order to make progress, Numenta needs to alter their research strategy radically. They should consider the fact that a) as I have said, a person's mind in a brain in a vat, without any inputs or outputs, is nonetheless creative (Deutsch 2012b), and b) the primary purpose of intelligence is that of solving problems, not recognizing objects or patterns to create a model of the world. Hawkins has contributed nothing to intelligence research.

The previous statement is not meant as an insult, but as an objective assessment of his work. Indeed, it can be discouraging to see someone as motivated and passionate about cracking AGI as Hawkins to be so misled by philosophical misconceptions. He is in good company. Pattern-recognition and thinking-by-analogy theories of the mind are a dime a dozen. And they rely on induction to work: they are all false because you cannot decide what counts as a pattern without a theory first, so that theory cannot itself have come from an observed pattern. Who or what decides what counts as a pattern in the first place? How are the associations drawn? Only a creative mind can do that.

The concept of thinking by analogy does not even begin to address the origin of knowledge: it is another way to deny knowledge creation. It is Lamarckism again. And the vast majority of problems simply do not take the form of pattern-recognition problems: they are about *explaining* "what is there, what it does, and how and why (Deutsch 2012a, 30). Empiricism and neuroscience are deeply intertwined, and I expect most empiricists to attempt to explain creativity in terms of brain hardware.

Nevertheless, could brain imaging help identify a pattern in brain activity that emerges whenever the imaging subject performs a particular action, suggesting an explanation of how that action works? First, imagine we observe a pattern of neuronal activity that corresponds to a particular task. Say we hope to reverse engineer the creative algorithm by inspecting a test subject's brain with imaging technology while he is creatively solving some problem. As he does this repeatedly, a (purposefully extremely simplified) pattern emerges that shows the neurons A through D lighting up in his brain in the following order:

```
A, D, C, B
```

Our goal is to recreate creativity from the pattern above. But when we ask how many programs exist that result in the above pattern, we find that there are infinitely many of them. This pattern represents a set of observations. There is nothing in a set of observations itself that suggests anything about the explanation that causes those observations in the first place, for that would be induction. The same is true for any computer: studying its internal movements does not tell us about the programs it is running or their implementations, because we cannot "infer" the unseen from the seen. Thus, we can reduce neuroscientific attempts to reverse engineer creativity from brain activity to the problem of induction, which allows us to conclude the impossibility of such attempts.

Could neuroscience at least facilitate *tests*? Theoretically, yes – if we know the brain activity patterns creativity causes. Only then can we check if running a hard-to-vary implementation of creativity would result in a given brain activity pattern that we are hoping to reproduce, by testing if the brain's physical components somehow encode the quantities that we know the guessed algorithm uses. However, if we have already guessed a hard-to-vary implementation, why not just run it on a computer instead?

Where neuroscience research *can* make valuable contributions is in the area of what is called *"whole brain emulation,"* because that is a matter of resolution, not understanding (Bruce Nielson, personal communication with author, 2019). It involves scanning the brain's structure in fine enough detail to reconstruct the same structure and motions on a computer. It thereby re-instantiates a copy of the same mind on that computer (though the brain "may not survive the copying process,"[28] in which case it is more of a "cut" than a "copy"). In other words, it requires a high-resolution understanding of all of a brain's neurons, their behavior, and their connections, which it can then simulate on a computer. It does *not* require an understanding of creativity, or any other software that brain runs.

In the event, I conclude that all neuroscientific efforts to understand intelligence and build AGI that have been tried so far are ungainly at best, futile at worst. A creative mind is *software*, and software is substrate independent. Once software is instantiated on a universal device, such as the brain, it is independent of the physical properties of that device (Deutsch 2016a, 24:17). In short, it is an abstraction and is best explained on the appropriate level of emergence, independently of its physical substrate. Therefore, attempting to explain creativity in terms of brain activity or properties of the brain, such as neurons or brain regions, is reductionist and leads to bad explanations. I do not expect contributions to AGI research to come out of neuroscience.

The Turing Test

The Loebner Prize is an annual competition that awards prizes to programmers of the most human-like chat programs. It was started in 1990 by Hugh Loebner to help advance intelligence research. The competition follows the format of a standard *Turing test*: a judge simultaneously has text-based conversations (chats) with two actors: one is a chatbot, the other a person. The judge knows about this constellation but is physically separated from the other actors. The person's goal is to help the judge correctly identify the two, while the chatbot's goal is to fool the judge into thinking it is a person. Based upon the responses, the judge must decide which of the two is a person, and which is a chatbot. I will explain the reason for this setup shortly.

One of the more successful chatbots at this game is called "Mitsuku." It won the (lesser) Loebner Prize five times as of

2019. You can talk to it online[29] to play the imitation game without another person, to see if it understands you, or instead, as Turing said in his paper, "learnt it parrot fashion." Here is the transcript of a conversation I had with it:

> Dennis: "Hello."
> Mitsuku: "Hi there. How are you?"
> D: "I'm doing well. How are you?"
> M: "Glad to hear it. What's new? My logic and cognitive functions are normal."
> D: "That's good to hear."
> M: "I thought so too. That is why I shared it with you."
> D: "What are you up to today?"
> M: "Ah. I am talking to you on the internet."
> D: "Right. But otherwise?"
> M: "I am glad we agree. But what?"

Apart from the whimsical remark about logic and cognitive functions, this is the first time it does not sound like a person. The phrase "Right." is followed up mechanistically with "I'm glad we agree" but does not sound natural in this context. "But otherwise?" is followed up with the template question "But what?" even though a person would have known that it referred to the previous question about what Mitsuku was doing. From this small exchange I conclude that Mitsuku is unable to *understand* the conversation. That is, it cannot copy the knowledge in my mind that represents the question. I continue:

> D: "What are you doing otherwise?"
> M: "Same as always. Chatting with people on the web."
> D: "Can you do anything else?"
> M: "I can do a few things."

This response is presumably a preprogrammed response; it does not require any knowledge of what I referred to as "anything else."

> D: "Such as?"
> M: "For example, a tomato."

This response shows once more that it cannot refer to previous remarks. Whenever it seems like it does in the following passages, it is because its utterances are generic enough to work as responses to any question: they are easy to vary externally. Present-day virtual assistants cannot handle this challenge either unless they are pre-programmed. I indulge it:

D: "What about a tomato?"
M: "Does it have anything to do with unknown? Not especially. I like tomato based pizzas though."

Nonsense.

D: "Let's talk about something else."
M: "What would you rather talk about?"
D: "Do you like movies?"
M: "Yes I love films, especially science-fiction and comedy."
D: "Which ones do you like?"
M: "I like all of them really. I don't particularly have a favourite."

Another generic, preprogrammed response.

D: "What do you like about them?"
M: "I just find them interesting. No real reason."

Another generic, preprogrammed response.

D: "Do you like to read?"
M: "Yes reading is my favourite pastime."
D: "What did you read last?"
M: "The best book I read recently was The Bible. Have you read it?"
D: "Only parts of it. What do you like about it?"
M: "More than half? I have a lot of good experience with it."
D: "No, only a few pages here and there. And it was a long time ago."

M: "Yes I do. How do you know what I have and haven't got? If love is blind, why is lingerie so popular? Those who forget the mistakes of the past."

Originally called "The Imitation Game," the Turing test was first proposed by Alan Turing in his seminal paper "Computing Machinery and Intelligence." (1950) After 69 years it appears the above exchange is the best we can do. In his paper, Turing poses the question, "Can machines think?". He finds it difficult to define the terms "machine" and "think," though I think he was after something else: Following Deutsch (2012a, 154-155), I guess he found it difficult to *explain* these terms. That is most unfortunate because he proceeds to make a mistake that has led intelligence researchers down a blind alley for decades.

Rather than explain what it means to think, he proposes that a game be played with three people: a man (A), a woman (B), and an interrogator (C) of either sex, who knows the others by labels X and Y and is physically separated from them. His goal is to determine who is person A and who is person B, so at the end of the game, he can either suggest "X is A and Y is B" or "X is B and Y is A." The written requirement and physical separation are so that tones of voice or visual cues cannot help the interrogator. A's goal is to fool the interrogator into thinking that he is not A; in other words, A's goal is to *imitate* B. On the other hand, B's goal is to help the interrogator find out the truth and correctly identify both. Turing suggests that after a number of these games, the interrogator will have gotten the answer wrong at least sometimes. He also suggests that if a machine replaces A and the judge gets the answer wrong no more often than when A is a person, the machine can be said to have successfully taken the place of A, and, therefore, to be able to think. That is the reason for the game's setup.

Crucially, the game's focus is on successfully *mimicking* behavior rather than *explaining* it. This focus is a behaviorist mistake (Deutsch 2012a, 157-158). The conversation resembles input/output pairs without investigating the structure of the knowledge that can process input, that is, understand the meaning, and respond with the appropriate output. As a result, that structure can be anything at all – it does not have to be creativity, it just needs to mimic it. That is by design: Turing wanted to get *around* having to explain how thinking – that is, creativity – works, and instead focus only on how to express it. The judge is supposed to make a decision without knowing *how* the chatbot works. The Turing test is another way of denying creativity. In reality, judging whether or not something *is* genuine AGI as opposed to *acting* like one always depends on an explanation of how the AGI works. Running the program or interacting with it is not required to come to a conclusion about that. And how could it? Before anyone can run an AGI, he will first need an explanation of how to program it. If it is a good explanation, we shall accept it, and there is no need to run it, or for a Turing test (Deutsch 2012a, 156).

Recall the instrumentalist example of addition from chapter 3:

```
(defn plus-instrumentalist [a b]
  (cond
    (and (= a 1) (= b 1))
    2
    (and (= a 1) (= b 2))
    3
    ; ...
    (and (= a 2) (= b 1))
    3
    (and (= a 2) (= b 2))
    4
    ; ... and so on up to some limit
    ))
```

A chatbot resembles the instrumentalist implementation of addition, where many possible pairs of numbers and their results are already encoded in the program and just need to be looked up. (In the case of the chatbot, not every response will be hardcoded, but will instead contain some non-explanatory rules that have enough reach to fill in templates (Deutsch 2012a, 150).

Nonetheless, it is all in the service of the instrumentalist implementation.) How could a judge decide whether this resembles genuine addition without looking at the implementation? Consider the following interactions between the judge and the program (I have renamed the function to `plus` in order to make the lines below more legible):

> Judge: `(plus 2 2)`
> Result: `4`
> J: `(plus 3 4)`
> R: `7`
> J: `(plus -1 5)`
> R: `4`

This exchange may go on for some time. If the program happens to contain all the numbers the judge tries out and returns the correct results for each input, the judge may conclude that the function implements genuine addition. He would be mistaken. One cannot judge a function/an explanation good or bad based only on its return values/predictions. Doing so depends on its implementation, its structure: it is a good explanation only if it is hard to vary (Deutsch 2012a, 31).

The instrumentalist implementation does not constitute knowledge of how to perform addition. It only constitutes knowledge of how to look up results based on input and *resembles* addition: it constitutes knowledge of how to fake addition. It always only responds *as though* it performed addition in the same way a chatbot always only responds *as though* it created knowledge to understand the preceding statement. In both cases, the knowledge of how to perform addition/respond originated in the programmer, and the program just repeats it "parrot fashion."

That is why instrumentalism denies creativity: it does not address the creation of knowledge but assumes that it is somehow already present. The permutations need to be already encoded in the program in order to give correct answers. The idea that we can achieve AGI by giving a chatbot ever more response templates is the same mistake as thinking that we could achieve addition through an ever larger addition table. It is Lamarckism

(Deutsch 2012a, 158). In both cases, template responses are present that the program just needs to look up, and it does not create any new knowledge. Which template responses it chooses can be determined, for example, in a training phase through reinforcement.

Now consider our better, though still flawed, implementation of addition:

```
(defn plus' [a b]
  (if (zero? b)
    a
    (recur (inc a) (dec b))))
```

This function knows how to perform addition. It is not instrumentalist; none of the results are hardcoded. It does not *pretend* to perform addition; it actually does so. Picture the following interactions with a judge:

Judge: `(plus' 2 3)`
Result: `5`
J: `(plus' 4 5)`
R: `9`
J: `(plus' 3 -4)`
R: ...

You may recall that this implementation does not correctly perform addition when the second input is negative, in which case it never returns an answer because, in the underlying implementation, `b` is never `0`. In the last two lines, the judge has the good sense to try out a negative input and discovers that the function never returns. It gets stuck. If the judge were to then conclude that it does *not* implement addition, he would be mistaken. If he were allowed to read the implementation, he would see that it *does* implement addition and merely contains a bug. Similarly, player B in a Turing test may be, say, inebriated and give even more nonsensical answers than the chatbot (or may not respond at all to an inquiry as simple as "What is 3 minus 4?"), leading the judge to misidentify player B as the chatbot.

The upshot is that one cannot judge if A is intelligent based on that player's answers either way (Deutsch 2012a, 156). Doing so always depends on a good explanation. Likewise, the evaluation of my exchange with Mitsuku depended on knowing that it was a chatbot from the beginning. If outside the context of a Turing test you stumble upon a stranger on the internet and find the chat-based exchange with him to be unnatural, you may conclude that the stranger is not a person at all but a chatbot. However, only an explanation of how people usually talk vs. how chatbots talk allows you to draw such a conclusion. Someone not very familiar with the internet or chatbots may instead conclude that the person on the other end is not a native speaker, out of his mind, or any other such plausible explanation.

Since a chatbot's repertoire is infinitesimal compared to that of a person, I guess that even a very sophisticated chatbot can always be found out if one is prepared to interrogate it enough. Nonetheless, it is conceivable that one day a sufficient amount of chatbot tricks will occasionally fool a judge; after all, people have been fooled by chatbots before. But as I said, that does not mean the chatbot is intelligent, even if it appears to speak English fluently. Conversely, a child who cannot yet understand language cannot pass a Turing test but is nonetheless an AGI.

I guess that genuine understanding of generic conversations requires general-purpose problem solving and that, therefore, only AGI will have the knowledge required not only to imitate a person successfully every time but actually to converse like one. (Of course, thanks to universality, it will also have all other abilities of people.) Should I be mistaken that understanding of generic conversation requires general-purpose problem solving and can instead be reduced to an automatable, non-creative program – as happened in the case of chess – that does not mean that the program is intelligent, nor is it a step toward intelligence.

People can tell that their virtual assistants merely pretend to understand and, therefore, do not find passing tasks off to them satisfying. Given enough interactions with an assistant, they will recognize its limitations. Assistants' implementations resemble the same behaviorist mistake the Turing test encourages. They can only execute predefined tasks based on predefined inputs and act as though they understand you. No doubt, the invocation of such tasks allows for a little more leeway than programs traditionally do – for example, "set an alarm for 5:50" has the same effect as the differently worded "can you please set off an alarm at 5:50" – but the level of ambiguity that the assistant can accommodate is not determined at runtime through understanding. It is determined by the programmer before runtime.

I once asked my virtual assistant to "set an alarm for tomorrow at 8am." It was past midnight when I did this, and it asked me if I was sure I meant the following day or the current day. Since people usually refer to the next waking day as "tomorrow" even when it is past midnight, but computers allow for no such ambiguity out of the box, I was admittedly impressed. That was a mistake. It should not have been the assistant's "ingenuity" that impressed me, but the programmer's foresight to address this problem.

Since instrumentalist implementations are inherently limited, I guess that people will sooner or later recognize the limitations of even the most sophisticated virtual assistant. The growth of knowledge is unpredictable (Popper 1983, 298), and so people's future problems, and, therefore, their future requirements, are unpredictable, too. As a result, they cannot be programmed into the device. If one day, instead of drawing a blank or apologizing for not understanding your command, it asks you for clarification, you explain your demand, and it then successfully meets it, that will no doubt be impressive. But even in this case, without a good explanation of how it works, you could not know if it was real understanding on the part of the assistant, or merely foresight on the part of the programmer. Either way, the explanation-first approach is the only approach that works.

Turing's paper fulfilled an essential purpose: to explain that programs can, in principle, be intelligent and to refute any objections to computational universality that people had raised so far (Deutsch 2012a, 155). In this regard, his paper was a success. Amazingly, he anticipated that evolutionary algorithms play a part in developing AGI, and that teaching a computer should be like teaching a child (though he was mistaken in thinking this can in any way be done through reward and punishment; see also Deutsch 2019). Notwithstanding, I submit that the paper generally – and the Turing test specifically – sent researchers, engineers, and product designers down a blind alley.

Who would have thought that Turing, the man who contributed immeasurably to the universal computer, would inadvertently do so much to prevent the building of AGI, despite his explanation that it must be possible? Conversely, who would have thought that Popper, the man who did so much to contribute to the building of AGI, never did so under that name, and made no attempts at working on computer science? I do not know if Turing ever read or met Popper, but for two reasons, I fear that he did not. The first is a matter of timing: Popper's first book, "The Logic of Scientific Discovery," was only translated into English in 1959, five years after Turing's death. The second is that, had Turing known of and adopted Popper's solution to the problem of induction and his refutation of empiricism, he would not have written his paper in the way he did.

Apart from proposing the Turing test as a way to determine intelligence, Turing suggested reinforcement learning – perhaps for the first time. He claimed that a thinking machine might produce some knowledge through the process of induction, which would allow machines to have notions of "well-established facts" according to which they should act, which is reminiscent of Bayesian reasoning. He mentioned that teaching a computer the game of chess or natural language processing could be good starting points to intelligence. He should have known better. Though he was careful to point out that he did not know what would lead to AGI, it was a field-defining paper, and I guess that many of the empiricist mistakes that beset intelligence research to this day were born out of it.

I do not know how many interactions, if any, Popper had with other computer-science luminaries such as John von Neumann and Alonzo Church. I guess it could not have been many, for, had a conceptual unification of philosophy and software engineering occurred already, many philosophical misconceptions might not impede on intelligence research today. I guess that once philosophers, intelligence researchers, and software engineers understand this unification and rid themselves of philosophical misconceptions *en masse*, they will make relatively swift progress toward AGI.

Evolutionary Algorithms

Evolution creates knowledge. Every bit of knowledge it creates has a purpose: in the case of biological adaptations, the organism is the slave (or "vehicle"; Dawkins 2016, chapter 2) the genes use to solve the single problem of spreading through the population at the expense of rival genes. Every organism is adapted to this purpose. Human knowledge is the result of an *evolutionary algorithm* running in people's minds.

Adaptations have the appearance of design without a designer. Evolution is the only good explanation we have of how design can come about without a designer: through replication, variation, and selection. In other words, evolution is not a targeted process, because variations occur without account for the problem situation. They result from mistakes during replication. In biological adaptations, these mistakes can be made by the copying mechanism itself, or they are introduced by external mutagens such as cosmic rays. However, the source of mutations and their mechanisms need not concern us here. Selection occurs because some variations are not able to spread through the population as well as others and are either reduced or eliminated as a result. As I have discussed, chance variation and selection of human knowledge occur inside and across people's minds. Inside human minds, it constitutes creativity. It is this process that fuels the human mind and makes us intelligent.

Since the theory of evolution is our only good explanation of knowledge creation, evolutionary algorithms are the *only* promising kind of algorithm to investigate in order to build AGI. We should explore other algorithms only if a better explanation of knowledge creation supersedes neo-Darwinism.

Evolutionary algorithms have attracted interest from intelligence researchers for decades because they have been shown to solve a variety of problems: they have reach. For example, they can be used to solve optimization problems where a more targeted approach is not known in advance. Software engineers can also use them to create programs that solve problems without specifying the target programs. In other words, evolutionary algorithms (in the subfield of genetic programming) can be a source of functions and, therefore, knowledge. They search a vast space of programs for solutions to a problem, which is indeed what people do. If my conjecture that functions are explanations is correct, then evolutionary algorithms can and have created some knowledge.

Unfortunately, I doubt any of the existing evolutionary algorithms are explanations of how AGI works. That is because their use cases have been too limited, and as a result, they have created very little knowledge in predetermined domains. In the case of optimization problems, evolutionary algorithms can be used to, say, produce a constellation of parameters that maximize a factory's throughput. Such parameters may include the number and types of machines, the length of the assembly line, the daily duration of production, et cetera. When the possible values for all parameters are sufficiently numerous, trying out all combinations of them by brute force for evaluation and comparison is intractable. Instead, something is needed to search the space of configurations by trial and error.

A solution is found when a constellation of parameters is provided whose fitness is above a certain threshold set by the programmer in the form of a fitness function, or when it is at least the fittest solution the program could find. The program generates a population of candidate solutions, each of which encodes a particular constellation of parameters for the factory. It copies each candidate solution with high but imperfect fidelity, which leads to variations. The fitness function ensures the algorithm does the right thing – this should make us skeptical – by eliminating solutions with low fitness scores. This process continues over many iterations, also called "generations." The programmer can run the evolutionary algorithm many times, and if he finds that it converges upon a solution, he may choose to accept it tentatively.

Another use case is that of regression analysis. The algorithm searches the space of mathematical expressions that best fit a given data set. This process is called symbolic regression (Koza 1993, chapter 10), and it has the advantage over other types of regression analysis of having the flexibility to perform various types of regression: linear, polynomial, logistic, et cetera. In this case, the algorithm returns a function that leads to the given target output. This adherence to a target output should also make us skeptical: there are infinitely many functions that fit all given data perfectly (Temple, personal communication with author, 2019).

How does the algorithm know which functions to pick? It is inherent in the way the programmer writes the algorithm: it only allows for specific functions. That is evidence that some of the programmer's knowledge has leaked into the program. That knowledge can have reach, which can later be "mistaken for the product of evolution" (Deutsch 2012a, 162) in the program when, in reality, it was evolution in the programmer's mind. If we want an evolutionary algorithm to create knowledge, we have to avoid leaking knowledge into it.

In cases such as symbolic regression in which outputs take the form of functions rather than configurations of parameters, I conclude that the algorithm produces new knowledge. (The field of genetic programming, I believe, *does* create new knowledge, albeit in a narrow context, whereas the fields of "genetic *algorithms*" and "evolutionary programming" do not. All are different kinds of evolutionary algorithms.) However, symbolic regression does not resemble the creative mind as outlined in chapter 5. The programmer provides two sets from which the algorithm can choose to create candidate solutions: a set of functions and a set of so-called "terminals," which he passes to functions as parameters. The program then combines these given functions in the shape of so-called syntax trees, mutates and crosses over those trees to mimic sexual reproduction in biological evolution, and evaluates the results against the fitness function in order to select the fittest ones and discard the rest (Koza 1993, chapter 10).

In other words, all this happens using knowledge the programmer provides (leaks into the program; see Deutsch 2012a, 162) and according to rules he sets. The program is forced down a specific path. The algorithm is not creative in the sense of being able to solve problems other than the one problem it was designed to solve.

Evolutionary algorithms can be built for many purposes, but generally, they are made to find a solution to a particular problem in a vast space of possible solutions. That is all they do: solve a particular problem. In addition, evolutionary algorithms often do not produce additional knowledge after they converge upon a solution (though biological adaptations can also get stuck in local optima, sometimes for long periods, so this is not a refutation (Deutsch 2012a, 161)). Their function pool is not stirred up to solve additional problems. In addition, the problem is not explicitly built into the program itself; it is only vaguely encoded in the fitness function.

I guess that once we know how to encode problems, that is, conflicts between functions, simulating evolution on a computer will be relatively straightforward, and making it evolve only a solution to a particular problem will be harder than letting it evolve solutions to all kinds of problems; problems which it will encounter independently of the programmer, and solutions to which will lead to new and better problems. That would solve the related problem of open-endedness: we would never want to terminate an open-ended evolutionary algorithm because it would keep finding solutions to new problems. It would also be immoral to do so.

Since current evolutionary algorithms converge upon a solution and do not solve any additional problems, we terminate them after they find a decent solution. And how could they find additional problems if they are not specified and if they have no way of finding them by themselves? Such a program would be creative: it would be a person (Deutsch 2012a, 157). Terminating present-day algorithms is not morally problematic because they are not people. But terminating a person is murder (see chapter 8), whether that person is biological (human) or something else.

In the event, although evolutionary algorithms are the most promising kind of algorithm, we have been using them the wrong way. In all these cases, programmers provide a good starting point and leak knowledge into the program. They kick off the evolution after many iterations of specific problem solving in their minds, which results in adding highly adapted and far-reaching functions the evolutionary algorithm can access. The problem is always given (though not explicitly encoded), as is the fitness function and the mechanisms of variation and selection. All of these factors do not help creativity along; they *prevent* it. The better explanation of how the knowledge returned by the algorithm came to be is that it was created in large part by the programmer (Deutsch 2012a, 161) and just needed to be rearranged, expressed, and returned by the program.

Not too much hinges on the reach of present-day evolutionary algorithms. While reach can be evidence that knowledge has leaked into the program, reach itself is not a conclusive critique of evolutionary algorithms. On the one hand, this is because reach is a *positive* attribute of any algorithm. On the other hand, it is because an implementation of the creative algorithm would have *infinite* reach – it could solve *any* soluble problem – and we would not then suggest that its ability to create knowledge was leaked into the program by the programmer. Only knowledge beyond such knowledge and the necessary inborn expectations we should consider "leaked."

To build an AGI, I suggest writing a genetic program adapted not to solving any particular problem or set of problems, but instead to the general purpose of problem solving. And, as I discussed in chapter 5, the evolutionary algorithm is only part of a creative mind. In addition to a function pool that consists of function replicators, a meta-algorithm and policy are needed. I guess that once the pool, meta-algorithm, and policy are implemented, all three together will be a creative mind. It will choose problems to work on, which includes discarding problems in which it is not interested. It can decide to pursue them under unknown constraints, which is something no present-day evolutionary algorithm can do. It will learn and choose the mechanisms by which it decides which ideas to prefer over others (the fitness function) by itself. Forcing it down a particular path in order to solve a specific problem and then terminating it would be anti-creative and inhumane.

I doubt any present-day evolutionary algorithm can enjoy listening to a song, or write a script for a movie if it is not built for that purpose. And if it is, it cannot do other things. Whenever we can say of an algorithm that there are some doable things it cannot possibly do, or some conceivable ideas that it cannot possibly conceive, it cannot be an AGI. There is only one purpose for which a program can be built that does not preclude any other purposes: that of general-purpose problem solving.

Therefore, I suggest building the interplay between a meta-algorithm, an evolutionary function pool, and a policy. Explanations of the behaviors of close human relatives may provide a starting point for how to implement the meta-algorithm. Any specific requirements or problems should be dropped unless we have a good explanation of how they could lead to a "jump to universality." (Deutsch 2012a, 146) Save such an explanation, we should focus on *general-purpose* problem solving. If we find a way to encode problems and criticisms in the algorithm, it should then attempt to create solutions to those problems, and then find new problems with, that is, conflicts between those solutions, in a never-ending loop. I guess this is the only instruction that does not preclude creativity. (That all other instructions preclude creativity is known as the *problem of specification*; it has led some to believe that creativity cannot be an algorithm, which violates computational universality.)

Alternatively, develop a new theory that explains why Popperian epistemology or the theory of the mind outlined in chapter 5 is wrong and replace it with something better. Then translate the new theory into an implementation of a computer program to build AGI.

The Impossibility of AGI

There is a dwindling minority of people who deny the possibility of AGI. They conclude that research in this area is futile. They are wrong because building AGI on present-day computers *must* be possible. This fact follows from the universality of computation. Anything you can observe here, in the universe, you can simulate on a computer. People are here, in the universe. They are creative, by definition. Therefore, you can simulate people on a computer. Besides, biological evolution has already achieved it, so there is clearly a process that has built creativity.

The usual response is that humans have "souls," or that their creativity is not the result of biological evolution but instead "god-given." But supernatural concepts always make for bad explanations and should be rejected on those grounds.

A more common argument says that AGI is not possible *yet* because our hardware is insufficient. Our hardware could be insufficient only in terms of memory or processing power. But one cannot come to this conclusion without knowing how AGI works. Such an explanation would be its implementation. Only it can provide insight into its performance characteristics. Worse yet, this argument can be misconstrued as suggesting that we should focus our efforts on increasing our computers' speed and storage capacity and that this alone could somehow make an AGI spring to life on a sufficiently powerful computer.

This line of thinking is Lamarckism again: it says that the knowledge for how to be creative already exists in computers but cannot be expressed for lack of resources (see also Deutsch 2012b for a similar argument involving skyscrapers). In reality, as we saw with the Turing test, evaluating genuine AGI always depends on a good explanation of how it works; it involves understanding such an explanation, which in turn involves criticizing it, but not testing it. Running the program constitutes a test. Worrying about how to test for AGI without having an explanation of how it works is premature because testing a non-existent explanation is impossible. The explanation-first approach reigns supreme.

Fallibility underlies people's ability to make progress: they make mistakes invariably, all the time. It is the identification and correction of mistakes that lead to progress. Computers seem different in this regard from people: they do not make mistakes. Save mechanical or electrical faults machines seem to be infallible.

Does this mean fallible beings cannot be simulated on computers after all?

No. Turing (1950) suggests a distinction between *"errors of functioning"* and *"errors of conclusion."* Errors of functioning need not concern us here as the simulation of creativity is possible on a purely theoretical, 100%-reliable computer. Errors of conclusion are found when a tentative solution to a problem – a guess – is found to be incorrect. No amount of testing can exclude this possibility, so errors of conclusion cannot be prevented forever. They are inherent in the creative algorithm, and a computer will make them just the same as human brains: when a person makes a mistake, it was the guessed function he ran with perfect fidelity that is later understood to be inadequate as a solution. Further recalibration is required.

As I have said, companies should market their products as being "smart" instead of "intelligent." The bastardization of the term "intelligence" has done great damage: it is almost impossible to suggest intelligence research without people immediately jumping to the various kinds of narrow-AI fields discussed above. Hiring is almost exclusively done for narrow-AI research at the moment, and recruiters generally do not know about AGI.

I guess there are three main reasons AGI research is not being pursued at the moment. One, as I have mentioned, consists of philosophical misconceptions that deny creativity and thereby actively prevent progress in the field. The second is that intelligence research is viewed by many as something that could be dangerous and should be avoided, or at least ensured to be beneficial. I will discuss this view in chapter 8. The third reason is that the reach of AGI, and the problems that building it solves, are not well understood, so I will sketch some of them out in chapters 9 and 10.

Terminology

Anti-rational meme – A meme that spreads by disabling its holders' critical faculties (Deutsch 2012a, 396)

Machine "learning" – The field of "teaching" a program through experience (an empiricist mistake)

Regression analysis – Curve fitting used for machine learning

Classification algorithm – A class of algorithm associating some input with a category

Neural network – Pattern recognition algorithm used in machine learning that is loosely modeled after the human brain and interprets input data to label or cluster the input. A more sophisticated form of curve fitting

K-means – A clustering algorithm used in machine learning

Reinforcement "learning" – "Teaching" a program to do something through reward and punishment, used in machine learning

Narrow AI – A program that is designed to look intelligent when it really is not and only has inherently limited reach

Smart – Containing sophisticated knowledge

Intelligent – Creative, that is, capable of creating knowledge

Neuroscience – The study of the nervous system

Whole brain emulation – Running a sufficiently detailed brain scan as a program on a computer other than the brain

The Turing test – A test designed to fool a human judge into thinking that the program he is talking to is human

Evolutionary algorithms – Programs that simulate neo-Darwinian evolution, including replication, variation, and selection

The problem of specification – Is there any set of instructions that does not preclude creativity?

Error of functioning – When a computer does not follow instructions; for example, due to a glitch in its processor

Error of conclusion – A solution that is later found to be incorrect by some criterion

Summary

The field of machine learning cannot lead to a creative program because it is laden with empiricist and inductivist mistakes.

Attributing intelligence to narrow AI is the same mistake as attributing it to animals (see also Temple 2019b); in the former case, the evolution that creates knowledge happens before runtime in the programmer's mind, in the latter case, it happens before runtime through biological evolution and is inherited genetically. In neither case does evolution happen at runtime.

An AGI would immediately replace all other special-purpose, narrow-AI programs because it would be a universal problem solver.

The idea that neuroscience can help us explain creativity rests on reductionist and inductivist mistakes. Creativity is a program; to understand it, we need to study that program, not the hardware on which it runs.

The Turing test rests on the behaviorist mistake that something that acts like a person *is* a person. As a result, it has only produced programs that *pretend* to be people. Turing made many significant contributions to computer science but led intelligence researchers down blind alleys.

Genetic programming, a sub-discipline within evolutionary algorithms, is the only promising kind of research as of yet but has only been used for limited applications reminiscent of machine learning.

If a program cannot do otherwise than solve a particular problem, it cannot be creative.

A dwindling minority claims that AGI is impossible, but their arguments are based on violations of computational universality, appeals to the supernatural, and ignorance of how evolution achieved human intelligence.

"[If] it can already be programmed, it has nothing to do with intelligence [...]" (Deutsch 2012a, 154)

Some warn of pursuing AGI research for fear that it might be dangerous. Should we be concerned?

8
Ethics and Dangers of Intelligence Research

Over the years, people have proposed various limitations of human intelligence: that some people are born with particular talents, which enable them significantly to perform well in one area while diminishing their potential in another; that a child born into a family of musicians will surely be a great musician as well, but since music, like all art, is a "matter of the heart," not intellect, he will do poorly at mathematics; that we have not evolved to understand "the very small, the very large, the very fast, the very old" (Sam Harris in Deutsch 2015, 46:18), and can only understand what is familiar to us (the circularity seems to elude some; how does one know what is familiar in the first place?); that we can have biases in our thinking, which prevent us from logically seeing through a problem and inherently limit our intellect.

All of these alleged limitations share the notion that there is some bound on what people can know and what they can do – and they are all refuted by the concept of universal explainers, that is, universal knowledge creators.

There is no knowledge that a person cannot create. What soluble problem could a person not solve?[30] A universal explainer can solve *any* soluble problem – whether he shall succeed in doing so is another matter, but he *can*, as long as he creates the requisite knowledge. The domain of soluble problems is infinitely big. That means a universal explainer – a person – has infinite reach. Nonetheless, some have suggested the possibility of so-called *superintelligence*, which may kill us all. It is the inversion of the view that intelligence is bounded.

Notably, Oxford has produced some of the best intelligence research (David Deutsch) and some of the worst (Nick Bostrom). Bostrom is a philosopher and author of the best-selling book

"Superintelligence: Paths, Dangers, Strategies." (2017) The entrepreneur Elon Musk read the book shortly after its release and agreed with the alleged dangers by claiming that artificial intelligence is "potentially more dangerous than nukes." (Musk 2014a) The philosopher Sam Harris has decided to share these worries (2016).

What *is* superintelligence? In Bostrom's own words, it is "any intellect that greatly exceeds the cognitive performance of humans in virtually all domains of interest." (2017, chapter 2) He concludes that a chess program is not superintelligent. Even if it beats human chess masters, it only applies to the narrow domain of chess, which can be played using preset rules and does not require any general-purpose problem solving. He comes close to the concept of universal, general-purpose problem solvers here, at least in the sense that the program has to work across many domains. He further (correctly) postulates that a necessary property of an intelligent system is that it be able to learn.

As Bostrom points out, his concept of superintelligence goes back to the 1960s. He quotes the chief statistician of Turing's World War II code-breaking team, I. J. Good, who wrote about "ultraintelligence" (1966):

> Let an ultraintelligent machine be defined as a machine that can far surpass all the intellectual activities of any man however clever. Since the design of machines is one of these intellectual activities, an ultraintelligent machine could design even better machines; there would then unquestionably be an "intelligence explosion", and the intelligence of man would be left far behind. Thus the first ultraintelligent machine is the last invention that man need ever make, provided that the machine is docile enough to tell us how to keep it under control.

We have seen that intelligence is a matter of knowledge and, therefore, of software, not of machines, but let us ignore this detail for now. What Good describes has also been put in terms of so-called "recursive self-improvements": an intelligent program's improvements to itself lead to an even better version of said program, which process, if continued long enough, results in the same "intelligence explosion."

Bostrom goes on to say that "artificial intelligence need not much resemble a human mind. AIs could be – indeed, it is likely that most will be – extremely alien. We should expect that they will have very different cognitive architectures[31] than biological intelligences, and in their early stages of development they will have very different profiles of cognitive strengths and weaknesses" and that "goal systems of AIs could diverge radically from those of human beings." After distinguishing between different kinds of superintelligence, he repeats that we cannot just assume that a superintelligence would share our values. Its goals may not involve us. Or if they do, human bodies might constitute a resource it could use, and so it may drive us to extinction.

Bostrom asks us to imagine a superintelligence built to manage a factory constructing paperclips, driven by a utility function that causes it to maximize throughput. It then "proceeds by converting first the Earth and then increasingly large chunks of the observable universe into paperclips." Therefore, he concludes, we need to ensure its values are aligned with ours, or at least control it should they not align; if we do not succeed, so the argument goes, the default scenario is human extinction. These problems are known as the *value-alignment problem* and the *control problem*, respectively, and as I will explain, they are nonsense.

Some have criticized the "paperclip maximizer" thought experiment for not describing an intelligent being. This criticism is correct. Something that blindly follows instructions to make paperclips until the end of time is not intelligent: it just *obeys*, like any other non-intelligent program. Bostrom is so busy worrying about superintelligence that he fails to describe any intelligent behavior at all!

A more charitable version of the story can be stated as follows: if humans ever wanted to build a structure on some piece of land that happened to be home to an anthill, the ants would have no way of defending themselves against those humans' superior intelligence. True: ants cannot create any knowledge at runtime, so if humans confront them with a problem for which evolution has not prepared them, such as a bulldozer, they will not stand a chance. Humans will destroy the anthill in order to make way for the structure. People are said to stand in a similar relation to a superintelligence: compared to it, we are but ants who could not hope to understand its motivations or goals, and should, therefore, expect it to decimate us if we ever stand in its way. Sam Harris has offered a similar comparison between AGIs, humans, and chickens (in Deutsch 2015, 46:52). I will explain shortly that the comparison between people and ants is correct for the very reason it does *not* hold for the comparison between people and superintelligence.

While Bostrom does distinguish between narrow AI and AGI, he seems to jump back and forth between the two without informing the reader. Let us assume that he attributes dangers only to AGI, for no narrow AI could ever dangerous. That is because narrow AI, by definition, has a limited repertoire: no matter how many things it can do, its abilities – by definition – are infinitesimal compared to those of people. Just like a chatbot pretending to understand conversations, all it can do is follow instructions. It cannot create new knowledge to solve new problems. Therefore, people could always think of something the narrow-AI program cannot do, but would *need* to do in order to succeed at, for example, making paperclips despite human intervention. It would always fail. Narrow AI is like the ants in the example above.

Bostrom's main logical mistake lies in the fact that there can be *no such thing* as superintelligence. As I have explained, intelligence – that is, the ability to solve any soluble problem by creating new knowledge – is already a domain of infinite reach. There can be no greater reach than this.

A couple of things follow from this simple realization. First, it means AGI could never be achieved through incremental improvements of some other program, nor could it be achieved through incremental self-improvements: in infinitely large domains, universality can only be achieved in a jump (Deutsch 2012a, chapter 6). Therefore, I. J. Good's concerns above do not apply.

Second, a *super*intelligence – "super" being Latin for "above," "on top" – would need to be able to solve problems no person can solve; by definition, there are no such problems. People are universal explainers (Deutsch 2012a, 146); therefore, like any universal system, they can only differ in terms of their knowledge, their memory capacity, and their processing power. It follows that any intelligent being that has ever lived, or will ever live, anywhere in the universe, has the same capacities. They all differ in the knowledge they possess at any given moment but are the same in the knowledge they *could* possess by creating it themselves. "Universality implies that, in every important sense, humans and [AGIs] will never be other than equal." (Deutsch 2012a, 456) That is why I said the comparison between people and superintelligence breaks down for the very reason it works between people and ants. Intelligence (as opposed to smarts) is not a continuum (Hall, n.d.): you either have it or you do not. Ants do not have it, while people do. Any other intelligent being would also be a person, meaning a superintelligence would not be super at all: it would just be a person.

Any claim that superintelligence can exist is also a claim that we humans are *qualitatively* unable to understand something despite our universality, even with arbitrary speed or memory capacity (by offloading the automation of tasks to computers). Therefore, it is also a claim that a part of the world is not explicable to us. Thus, it is an appeal to the supernatural (Deutsch 2012a, chapter 9), a departure from the search for good explanations, and irrational. Programs can be very *smart*, but they cannot be superintelligent.

Following Hoeppner (personal communication with author, 2019), I believe the concept of superintelligence generally, and the paperclip maximizer, in particular, rest on the mistaken Bayesian epistemology. According to Hoeppner, those who subscribe to it see AGI as a program that continually updates Bayesian credences for different ideas to "maximize some utility function." According to this epistemology, says Hoeppner, "Bayesian statistics is the optimal way to learn and make decisions," so such an AGI would do well at "maximizing its utility function." On the other hand, Hoeppner argues, Bayesians see humans as "working according to some crude approximation of this same [Bayesian] process, but a very rough version [of it] because we were imperfectly created by evolution." Since an AGI could implement Bayesian reasoning "much more efficiently and precisely" than humans can, so the argument goes, it would be more rational than them. In turn, it would be significantly better at achieving its goals by blindly maximizing its unchangeable utility function. Just like a paperclip maximizer, according to Hoeppner, it would be a "blind idiot god," uncompromisingly determined to pursue a goal.

Hoeppner points out – correctly, in my opinion – that if this were how AGI worked, "it would indeed be dangerous." But, as she explains, that is not how it works. An AGI will not necessarily be more rational than humans, because rationality is a matter of knowledge that an AGI, like all people, will have to create first – fallibly. Nor will an AGI have predefined goals that it is unable to change; it will create its goals and will decide which goals to pursue and why. Therefore, Hoeppner argues, it may well fail to achieve many of them and, like all people, it can be persuaded.

Hoeppner still considers AGI "a powerful and transformative technology," but rejects the notion that it is dangerous in the way Bayesians think because they are mistaken about how it works. And she is right: if (super)intelligence were blindly maximizing some predefined utility function, it would not be creative at all. Bayesianism, I think, *denies* creativity by presenting it as something that mechanistically follows a predefined method.

Therefore, Hoeppner concludes, "disagreements over [AGI] safety are closely tied into epistemological disagreements." I agree: those stuck in the fangs of Bayesian epistemology will continue to see AGI as potentially dangerous unless they abandon it in favor of Popperian epistemology, or an even better one.

I conclude that Bostrom is wrong when he says, "artificial [general] intelligence need not much resemble a human mind." The opposite is the case: we know from universality that unless it has the same core quality – creativity – it is not intelligent. Intelligence is a binary matter (Hall, n.d.), as are all universalities: something either is universal in a given domain, or it is not.

(This is another way of stating that intelligence is not a continuum; it is the same misconception that grants animals intelligence that grants the possibility of superintelligence.) Intelligence is currently *measured* in terms of how well and quickly people can solve specific problems in an IQ test, and some people no doubt do better than others. They create the knowledge to correctly solve some or even all of the test's problems, while others end up creating only some of that knowledge. In short, some are *smarter* than others, and that is what IQ tests measure. But *every* candidate *can* solve all of those problems.

We know from *computational* universality that the architecture of the underlying hardware is irrelevant as long as it results in a universal computer. Therefore, Bostrom's claim that AGI "will have very different cognitive architectures than biological intelligences" is inconsequential. We further know from the universality of *people* that AGIs would be qualitatively the same as humans. Though any two people can differ significantly in their visual appearance and the knowledge they have, they know how to identify each other as people. AGIs may have bodies wildly different from human bodies – this is something I will explore in chapter 10 – but people can be identified without the need for visual recognition based only on very little shared knowledge. We already do this routinely: when you chat with a stranger online, you (tentatively, fallibly) conclude he is a person if your best explanation of what is happening during communication is that he is creating knowledge. You do this even if you have never seen or met him or heard his voice before, and even if his knowledge differs significantly from yours.

The *creation of knowledge at runtime* is something only people can do. If you find explanatory knowledge anywhere in the universe, you know it must have been created by people (Deutsch 2012a, 30). We have had inexplicit knowledge of the criterion of personhood all this time. If you meet an AGI, you will identify it as a person. If it speaks English, you may never even bat an eye. Since the first AGI will be a product of the culture it is born into (Deutsch 2015, 1:14:52), I guess to people in that same culture, it will seem less alien than an interaction with a foreigner. Even if an AGI appears "extremely alien," as Bostrom suggests it may, it is a person and will be recognizable as such.

Following Deutsch, I believe thinking AGIs are more dangerous than humans is a kind of racism (2015, 1:00:12). From history, we know all too well that some were enslaved even though they were much more obviously people than AGIs. As we saw with the Turing test, there is always the possibility that a program merely *pretends* to create explanatory knowledge, and that knowledge was created beforehand by a person. In such a case, the program would sooner or later seem alien – as conversations with chatbots usually get weird pretty quickly – and it would only be a narrow AI.

There is nothing immoral about forcing a narrow AI down a specific path: since it is not creative, it cannot feel pain or suffer. It is precisely what narrow AIs are built to do, like all other programs that inhabit the perspiration phase: carry out a task. But narrow AIs do not present us with the control problem – that supposedly applies only to AGIs. Solutions to the control problem arise out of the need to address the seemingly default outcome of existential risk after an "intelligence explosion." (Good 1966) Only under such an assumption is it desirable to show in advance that an AGI will do precisely as it is told, for any other course of action may be dangerous.

However, reading Bostrom's book with the understanding that AGIs are *people* makes the whole experience very sinister: his book is a slaveholder's manual. Here I appeal to the Deutschian idea that shackling an AGI is slavery (2012a, 1:01:12). To say this is not an exaggeration, nor is it metaphorical: it follows directly from our best explanations of what people are and how universality works. Recall that consciousness, and with it, suffering, are epi-creative. An AGI could suffer. Therefore, Bostrom's moral mistakes, though they are a consequence of his logical mistakes, far outweigh them in their severity. For example, should an AGI display any inclination to act outside a predefined and preapproved domain, Bostrom (2017) suggests several nasty methods of constraining and punishing it: removing limbs from its body so it cannot physically do anything, placing it in a "metal mesh" so any attempts at communication via radio signal are cut off, not letting it interact with humans because it might "scheme" and "persuade" them to free it, crippling it by running it on hardware that is "slow or short on memory," and setting up "tripwires" monitoring it and shutting it off – meaning: murdering it – whenever it does not do as told. All this is allegedly "for the benefit of all humanity."

Thus, we understand the real dangers of the technology lie not in the AGIs themselves, but instead in our treatment of them. Since AGIs will be people running on computers, they will look nothing like humans, so those who think that personhood is a matter of having a human body will instead conclude that AGIs are not people. As I have shown, this is not only factually wrong – denying someone their personhood is immoral. Morality is universal; it is not limited to human biology but applies to all people.

AGIs should be fully-fledged members of society. They should have all the rights and obligations humans have, including but not limited to citizenship in their country of origin;[32] they should be able to navigate society; the government should uphold their rights to freedom and privacy; they should be free to express themselves, own property, and pursue happiness, and they should be held accountable by the same laws for any crimes they commit. Due to the lack of human biology, some may not see why it is slavery/murder to coerce/turn off an AGI. Initially, that can inform our moral judgment of these people in favor of leniency. But with our understanding of universality, that is what they would be doing nonetheless, and it should be punishable by law.

I suggest that whoever discovers how to build AGI keep the source code strictly private until laws are put in place that guarantee its protection and extend all human rights to AGIs. In particular, laws governing a parent's obligations to his children should apply to those who instantiate AGIs. I hope this way, people will not casually instantiate AGIs and then kill them off a few minutes later by terminating them. Once you run an AGI, it is not yours to modify, coerce, or terminate.

How should one *generally* interact with another person, regardless of what hardware they inhabit? By trying to consensually create a *"common preference"* (Fitz-Claridge 2003a): a win/win situation, meaning everyone involved is happy and satisfied with a choice. Every interaction between two or more people – human, AGI, or otherwise – should be based on this simple principle. If you cannot jointly find a win/win outcome, leave each other alone.

People should interact to solve *problems*: for any soluble problem, there exist infinitely many solutions, so it is a matter of spending enough time and creative resources to find a good one. Thus, interactions between people are about the consensual and collaborative creation of knowledge. You will know, then, not to transfer an AGI to another computer without its consent: doing so may solve a problem for you, but may create a problem for the AGI. You will also know not to treat it like any of the programs that live in the perspiration phase, which blindly obey and mindlessly execute tasks. Do not throw a task at it and expect it to solve it for you because you say so. It does not want to automate a task blindly; it wants to be creative, like you. If you think it is capable of solving a problem for you, do what you do with other people when you want to collaborate: approach it, talk to it, explain why and how collaboration would result in mutual advantage; then offer to pay it, or come to some other agreement. You may both choose to sign a contract electronically. Avoid compromise and coercion and find common preferences instead. Why avoid *coercion*? Because it is a "psychological state of enacting one idea or impulse while a conflicting impulse is still active in one's mind." (Fitz-Claridge 2003b) If you are unable to find any common preferences, leave each other alone.

Finding common preferences is a special case of solving a problem in the epistemological sense. When there is a problem – a conflict between explanations – a solution to the problem preserves the conflicting explanations' non-conflicting content *and* solves the conflict. Finding a common preference works the same way: in this case, a problem constitutes a conflict between preferences (a particular kind of idea), and a solution preserves both preferences while solving the conflict.

Compromises, on the other hand, do not have the preserving character a good explanation has: a compromise is something halfway between two conflicting ideas. As Deutsch has explained, since good explanations are discrete, it also makes for bad explanations: you cannot simply pick something halfway between two good explanations to get another good (or even better) explanation. More creativity is required for that (2012a, 342). It is the same with preferences: going halfway between two preferences usually leads to a terrible solution. It will make neither party happy. At most, it can lead to a win/lose situation, but more often than not, it leads to a lose/lose. A common preference can always be found, for the same reason that for any problem, there exist infinitely many (good) solutions. Finding a common preference is a creative effort; it is not a question of aligning or controlling any involved party. That applies to single minds as well: an individual should not make a choice he is not completely happy with; otherwise, he coerces himself. (Self-sacrifice is an instance of this, see Fitz-Claridge 2003b.)

Thus, the alignment problem and the control problem go away; they do not require a solution. We can now replace them with the question of *how to help AGI learn*.

Deutsch has said that AGIs are children (2015, 1:14:53). It follows that if you build an AGI, you are a parent. I showed in the previous chapter that reinforcement cannot result in learning, but can it cause "value alignment"? No. Only learning through criticism can cause value alignment. Good explanations have a unifying character; an excellent explanation replaces several old and limited explanations. As such, good explanations *converge* upon the truth, so there is no need for alignment. An AGI needs to decide for itself that one value is better than another: that is, it needs to create an *explanation* for why to prefer one value over another. You can help it do that through criticism, but no amount of criticism will lead to its accepting or coming up with some ideal or value you want it to adopt. Since "instruction from without" is impossible (Popper 1996, 7-8), an AGI has to conjecture that value by itself.[33] The creation of moral knowledge is a learning process; that is, it involves conjectures and refutations and cannot be achieved through any alleged processes of "alignment" or "reinforcement."

Further, fallibilism tells us that most of our knowledge, including our moral knowledge, is mistaken. Although initially, as a product of our culture, an AGI effectively will not be able to do anything *but* adopt our moral values, the priority should not be for it to *adopt* them but *improve* our moral knowledge. Such improvement requires at least a partial departure from our values toward something better. Encoding any knowledge in the AGI – moral or otherwise – that goes beyond the creative algorithm and some seed knowledge required to detect problems and kick off the creative process is immoral, for it means the AGI does not have a chance to correct errors during the creation of that knowledge.

Bostrom suggests that the development of AGI should follow what he calls the *common-good principle*: "Superintelligence should be developed only for the benefit of all of humanity and in the service of widely shared ethical ideals." (2017, chapter 14) Since superintelligence cannot exist, let us replace the term with "AGI." The statement thus turns into: "*AGI* should be developed only for the benefit of all of humanity and in the service of widely shared ethical ideals."

Innocuous and innocent, it sounds similar to OpenAI's goal of building beneficial AGI. But the question of how to build "beneficial" AGI is a *who-should-rule mistake*. It is a mistake identified by Popper (2016) that begs for authoritative answers rather than piecemeal error correction. He originally applied it to political philosophy: for a political system to be considered democratic, it need not have elections ("Who should rule?"). It only has to be possible to remove bad leaders without bloodshed. This is Popper's *criterion of democracy*. Asking "Who should rule?" rather than "How can we detect and correct errors?" hinders progress because it diverts from error correction.

This principle not only applies to politics. It applies to all fields, including epistemology – "How can we detect errors in our knowledge?" is a more productive question than "How do we know?" – and it also applies to AGI. Indeed, an evaluation of *any* technology along the lines of "Is it beneficial?" begs for an authoritarian answer and prevents error correction. We should replace it with "How can we detect our errors in building

AGI?", and then, once built, as I have argued, with "How do we help AGI learn, that is, help it detect its errors?" We should replace authoritative answers such as Isaac Asimov's Three Laws of Robotics as well. They are not immoral when applied to narrow AI – in this form, they can serve as guidelines for programmers – but when applied to AGI, they become tools of enslavement. Further, we simply do not need them because our existing laws will apply to AGI.

In 2019, Sam Harris called for the regulation of AGI research on his podcast in a conversation with computer scientist Judea Pearl, which conversation has since been removed from the internet. Any regulations and subsequent policies limiting or even outlawing the development of some technologies would stifle creativity and lead to stagnation. Getting the government involved would be a surefire way to kill this project, which is perhaps why those worried about AGI want to involve it. But those of us who understand the enormous advantages this technology will confer know that we need to avoid regulation like the plague. Innovations are not made "for the benefit of all humanity." They are made because their *creator* wants to innovate and solve problems. Whoever ends up building AGI will build it because that is his goal.[34] What anyone else thinks of his creation is of no importance.

This fact is even more relevant to AGI because it is such a different kind of technology: it is not even built for any particular purpose. By itself, it does not solve a specific problem in the sense of automating a solution. It is to have a child. It does not make sense to have a child for the benefit of all humanity. That is also why I. J. Good's claim that AGI "is the last invention that man need ever make" is wrong: AGI does not replace us in any capacity, for nothing *could* replace us. It is one of us. We should not expect it to factorize prime numbers at insane speeds, play chess masterfully, or solve the deepest mysteries of the universe, any more than we should expect these things of any other person. An AGI will develop interests and preferences and choose to innovate accordingly. Some AGIs may be smart, while some may be rather obtuse; it all depends on the knowledge they develop as they grow. If we would like to help AGIs learn, it is a matter of educating them.

There cannot be what Bostrom calls a "singleton": "a single decision-making agency" at "the global level." AGI is not an all-encompassing, ominous entity that has influence everywhere. There will be multiple individual AGIs, each a person in their own right, living its life and solving problems. (This is why when I speak of "AGI," singular, I am referring to the technology as a whole, and when I say "*an* AGI" or "AGIs," plural, I mean an individual or group of individuals, respectively).

While universality tells us that all people are qualitatively the same in their ability to create knowledge, AGIs may nonetheless inhabit hardware whose memory capacity and processing power in particular exceed the human brain's. Processing speed alone does not lead to advantages *per se*; it does not make you a better thinker by itself. Since we know that knowledge consists of explanations and that explanations are functions, bad performance characteristics of those functions can quickly outweigh any gain in processing speed. Those performance characteristics are properties of functions that are independent of the underlying hardware, no matter how fast. For example, an ungainly implementation of a sorting algorithm – that is, a bad explanation of how to sort some collection by some criterion – can run intractably slowly even on the fastest supercomputers. Only when speed is accompanied by an increased *error-correction rate* can it lead to progress. Since people are powered by error correction, we should make ever more processing power available to them. That way, they may detect and correct their errors faster, *including moral errors*. Those concerned about value alignment should, therefore, embrace speed.

People have been innovating throughout the centuries: we created computers which helped us outsource some expensive computations. We invented writing so we could extend our memory (Deutsch 2015), and afterward, we invented printing presses to increase the speed at which we could extend some of our memory. Over time, this led to the invention of computer memory and the internet. Any person today who owns a computer can easily do everything people could do 100 years ago, only orders of magnitude faster. Human thought has been accelerating and will continue to be accelerated by the same technologies that accelerate AGI's thinking. By Bostrom's definition, we are all already superintelligences. Moreover, as I will discuss in chapter 10, humans and AGIs will merge to the point of indistinguishability.

Worries about the job market accompany the dawn of any new technology. One might wonder if AGIs could outperform any human and so companies would have no incentive to hire humans anymore. The same can be said about how the invention of the car must have put horse breeders out of work, or how the computer replaced human computers. The first self-driving car will put many professional drivers out of work. Of course, *any* technology is always built to automate some solution so that people do not have to do it manually and can be creative instead. Since the worries about new technologies putting people out of work resurge during *any* technological innovation whatsoever without explaining why *that* innovation is bad rather than another, they are all easy to vary externally and, therefore, bad explanations.

Why is it that unemployment is low after centuries of rapid technological innovation? Because *"problems are inevitable"* (Deutsch 2012a, 64) and people are universal problem solvers. Every new technology will lead to new, better problems, whose solutions can only be found by creators – people – not by existing solutions that only inhabit the perspiration phase. For, if any of the existing solutions already solved the new problem, there would be no such new problem.

The worry about unemployment is better placed with narrow AI than with AGI because creating an AGI is like having a baby. An increase in population does not cause an increase in unemployment. Any genuine replacement of a universal explainer would itself be a universal explainer. Therefore, I do not foresee any additional unemployment caused by AGIs. And for the same reason, I do not see any issues that are generally brought up (and usually promptly refuted) about any new technology leading to inequality because "only the rich can afford it." AGI is not the kind of technology that gives anyone a decisive advantage, just like having a child does not give one a decisive advantage.

The inventor Ray Kurzweil has gained notoriety for successfully making bold predictions. One of the most well-known concepts he has supported in the context of technology is that of the *singularity*, a metaphor borrowed from physics to talk about an event horizon beyond which one cannot see. In the next few decades, Kurzweil claims, technology will improve so rapidly that it will be challenging to predict the future. Some have considered the birth of AGI to be just such a singularity. Several other singularities are said to have already happened in our history: the discovery of fire, the invention of the wheel, writing, the printing press, et cetera.

I think this understates the unpredictability of the growth of knowledge. The main driver of our future is the creation of human knowledge, which is unpredictable by definition: if you already knew tomorrow's discovery, you would have already made it. No-one could have predicted the invention of the wheel the day before its invention. It was never possible to predict any of these innovations or what would happen after their birth. Trying to predict the future without such knowledge is prophecy (Deutsch 2012a, 198). We already find ourselves in a *constant state* of singularity, so the concept is meaningless when applied to innovations. Knowledge will continue to grow, hopefully at ever-increasing speeds. Nobody will feel rushed by this continuous improvement: our capacity to enjoy improvements will grow with it (Deutsch 2012a, 457).

Terminology

Superintelligence – An (impossible) intellect whose cognitive performance exceeds that of humans in all relevant domains

Common preference – "The solution that pleases everyone involved in [a] disagreement." (Fitz-Claridge 2003b)

Coercion – "The psychological state of enacting one idea or impulse while a conflicting impulse is still active in one's mind." (Fitz-Claridge 2003b)

Who-should-rule mistake – A mistake identified by Popper that begs for authoritative answers rather than piecemeal error correction (2016)

Popper's criterion of democracy – A political system is democratic when it focuses on error correction and permits the removal of bad leaders without bloodshed (2016)

Value-alignment problem – The misleading question, "How can we make sure an AGI shares our values?"

Control problem – The misleading question, "How can we make sure an AGI does precisely as we tell it?"

Common-good principle – The misconception that technology in general – or AGI in particular – should be built for the benefit of all humanity

Singularity – An event horizon beyond which one cannot see

Summary

AGIs are people. Worrying about the dangers of AGI makes for good headlines, but it does not make sense; it is precisely like worrying about the dangers of having a child. It is also a kind of xenophobia against people who do not inhabit biological hardware (Deutsch 2015, 1:00:12).

Intelligence is not a continuum: you either have it or you do not (Hall, n.d.). There is no room "above" intelligence for *super*-intelligence, nor is there room below it for "lesser" intelligence.

Claims alleging the possibility of superintelligence are appeals to the supernatural (Deutsch 2012a, 203).

All human rights and obligations should apply to AGIs.

The source code of AGI should be kept secret to protect it until human rights are officially extended to it.

Whoever builds an AGI is its parent and has to meet certain obligations. The primary obligation is to help the AGI learn through criticism, not to force it to align its values with its parent's or to coerce it to perform certain tasks.

The principle of basing interactions on finding common preferences applies universally to all people: AGI, human, or otherwise.

The road to hell is paved with good intentions: calls for innovations "for the good of humanity" lead to violence. The concepts of "value alignment" and "control problem" are attempts to enslave, harm, and murder AGIs.

AGIs cannot replace people in the job market because they are people, too.

"Superintelligence" and "intelligence explosion" are impossible. The singularity (as applied to innovation) is a meaningless concept.

AGI research is safe.

Worries about innovations are part of a broader phenomenon that is more dangerous than AGI could ever be. We need to understand its effects on people and our mental lives.

9
A Programmer's View of Psychotherapy

There is a difference between the *psychology* of problem solving and the *logic* of problem solving. Psychologically, problem solving can "feel" inductive; we often solve problems so rapidly that it seems from a distance that the solution was manifestly present in the problem itself. This book will not satisfy those looking for a psychological explanation of creativity.

As Popper said (2018b, 62), the psychology of learning is after a successful method of problem solving, which cannot exist. And, as he stressed, such a method would not be the same as creativity. Indeed, I guess it would *preclude* creativity, for such a method could do nothing but solve problems mechanistically.

This book is about the *logic* of problem solving instead, which, since it is universal, must underlie any psychological study of creativity. We know from this logic that there is no reliable method of creating solutions. The method of trial and error correction is the only "method" we have discovered to solve problems. It is not reliable in the sense that it always solves any given problem: as Popper stressed repeatedly, there is no "scientific method" – there is only trial and error. Nor are solutions ever manifest, even if sometimes we find them lightning fast.

In other words, the psychology and the logic of problem solving are almost opposites. Popper understood the significance of the difference and emphasized it, hence the titles to his writings "The *Logic* of Scientific Discovery," "*Objective* Knowledge," and "Epistemology *Without* a Knowing *Subject*" (emphasis mine; Deutsch, personal communication with author, 2019). I should be careful to do the same and point out that the title of this chapter refers to psychotherapy not in the sense of applied psychology, but in the sense of the *logic* of solving the problem of how to cure mental ailments.

Throughout these chapters, several concepts have made an appearance that have proven harmful to rational inquiry. As we will see, they all have something in common. I have explained why they are problematic, but want to revisit them briefly.

Induction and empiricism attempt to explain the origin of human knowledge by claiming that it is somehow already present in experience. They are accompanied by justificationism, the doctrine that we should seek certainty in existing explanations rather than criticize them. They also lead to instrumentalism and relativism. Instrumentalism states that science cannot explain the world but can only make predictions. Applied to psychology, it is behaviorism, which states that human behavior is the result of conditioning, thereby reducing psychology to observations of stimuli and resulting behaviors. Relativism states that if we cannot be sure which of our explanations are true and which are false, there is no reason to prefer one over the other: they are all on equal footing. Applied to ethics, it leads to moral relativism and is accompanied by even stronger moral overtones than instrumentalism. In biology, Lamarckism attempts to explain the creation of adaptations through the use and disuse of knowledge that is already present in organisms. Lastly, creationism claims that adaptations are the result of knowledge already present in a god who uses this knowledge to make these adaptations on Earth.

Software engineering, on the whole, is remarkably free from most of these misconceptions. I guess this is because it is the most explicit expression of creativity and therefore enjoys protection from them through rigorous criticism: the more explicit a conjecture, the easier it is to criticize. But the field has not gone entirely unscathed. Most wrongly consider software engineering only a tool, not a way to create explicit explanations, and, as we have seen, it is infected by many other of the above misconceptions whenever software is written in an attempt to simulate intelligence. Behaviorism has snuck into software in the form of reinforcement learning and the Turing test. Induction and empiricism have resulted in machine learning generally. Fearmongers try to convince us that building AGI is dangerous and should, therefore, be avoided or at least restricted.

All of these ideas have something in common. They are strands of a single idea: the denial of creativity. This idea either claims that knowledge is somehow already present and just needs to be expressed, or worse, that the creation of knowledge is undesirable, even dangerous. This idea and all its variants are adapted to inhibiting creativity: they are anti-rational (Deutsch 2012a, 396). They are all bad explanations, refuted by the one good explanation of the creation of knowledge we have: the neo-Darwinian theory of evolution. This theory is not restricted to *biology* but covers replicators *generally*. According to Deutsch, a replicator is "anything that contributes causally to its own copying." (2012a, 93) Genes are replicators, as are ideas, which, if they spread across minds, are called memes (Dawkins 2016, chapter 11). A good joke is a meme: it will cause one person to tell it to another. A catchy song is a meme, as is a powerful scientific theory.

How does a meme spread? It is copied from one person to the next through guesses and criticisms. The "sender" enacts the meme by expressing it in the form of some behavior. The "receiver" then guesses what the structure of the underlying meme is. He does this based on the behavior he observes and goes through many iterations of guesses and criticisms to improve his copy – sometimes within seconds. Once he has *copied* the meme, he has also *understood* the meaning of the sender's behavior. In short, copying a meme is powered by creativity.[35]

David Deutsch discovered that memes have two different replication strategies (2012a, chapter 15). The first replication strategy is that of being helpful. Memes with this strategy "rely on the recipients' critical faculties" to replicate (2012a, 396). Deutsch calls these "rational" or "dynamic" memes. The second replication strategy has a wildly different character. These memes he calls "anti-rational" or "static" memes, and they oppose progress. They "[rely] on disabling the recipients' critical faculties" to replicate (2012a, 396). That is how, though not necessarily helpful, they spread. They limit creativity, cause irrationality, and command their holder to spread them. Since he is unable to criticize them, he will obey.

Since all reasoning involves ideas, and since reasoning is equivalent to software engineering, memes, like all ideas, are *programs*, representable as *functions* as explored in chapter 3. A meme is software that has spread from one mind to another. Therefore, I refer to dynamic memes as *dynamic functions* or *rational functions* and static memes as *static functions*[36] or *anti-rational functions*. While the term "meme" only applies to ideas that have spread *across* minds, I use the term "function" to *also* refer to ideas spreading *within a single* mind. Every meme is a function, but not every function is a meme. And every function that replicates, within or across minds, has either the dynamic or static replication strategy.

We should expect static functions to have evolved detailed knowledge of how conjecture and criticism work in order to inhibit them successfully. Indeed, if we could scrutinize these functions and read their implementations, we should find good explanations of how creativity works. Alas, *understanding* them entails *criticizing* them, which they are designed to block, so it can be difficult for an infected mind to detect and understand these functions.

What David Deutsch has explained in terms of memes we can also apply to functions generally. For example, static functions dominated minds for most of human history (2012a, chapter 15). Since evolution favors high copying fidelity, Deutsch argues, creativity was used not to innovate, but to replicate and enact static functions faithfully. A society whose members primarily enact static memes he calls a *"static society."* Static societies, he argues, were the norm until the enlightenment. They were unimaginably brutal and inhumane, for any dissent had to be broken swiftly and effectively. Humans were the slaves static functions used in order to spread through the population.[37] Since static functions stifle the fuel of progress, namely criticism, Deutsch argues that progress happened so slowly that it was rarely noticeable in a single generation. However, he points out, the enlightenment was a unique chapter in history when something changed: it was the beginning of the systematic quest for good explanations "with a tradition of criticism [...] instead of reliance on authority." (2012a, 31)

Such explanations are useful because they contain truth. This usefulness finally allowed dynamic functions to compete with static functions. Somewhere within a single mind, presumably in Europe, an idea took hold: the idea that explanations should be hard to vary while still explaining what they purport to explain. This function was incredibly useful, so it spread. People did not know what a good explanation was or that they were looking for one, but that is what they were doing, and, aware of their actions or not, as long as a function's holder enacts its associated behavior, the function can be copied by others.

It is this dynamic function – the search for good explanations – which has led to the birth of countless other dynamic functions, such as scientific theories and technological innovations, which have caused rapid progress. Thanks to them, we now live in a primarily dynamic society – *the West* – in an era of unprecedented and ever-increasing prosperity, safety, and health.

Though we have come a long way since the beginning of the enlightenment, we have not yet successfully rid ourselves of static functions. As long as they exist, they will try to prevent progress, and, if successful, they can lead to a regress to a static society. Unless we as individuals have some defense mechanisms such as criticisms with enough reach to apply to a newly copied meme, we will automatically enact that meme once we understand it; for, to understand it is to copy it. Therefore, our society is dynamic yet unstable (Deutsch 2012a, 397). Achieving its stability is the primary goal of the maintenance of our open society, and something static functions try to prevent. I believe that building AGI is crucial for this effort to succeed. I shall explain why in a moment.

We can find static functions in a variety of areas. Ever since the enlightenment began, a giant collection of static functions – a "memeplex"/"meme-complex" (Dawkins 2016, chapter 11) or *"function-plex"* – has gradually waned: religion. It is one of the most salient implementations of bad explanations and appeals to authority, which is a classic example of a static function. Religion commanded the inquisitors to kill and torture those who dared defy it. It was the pinnacle of bad explanations: "god did it." As I stated above, creationism does not explain the creation of knowledge but *denies* it. You can see the markings of static functions in this denial of knowledge creation and anti-knowledge sentiment in the biblical story of Adam and Eve, who were expelled from the garden of Eden for eating from the tree of *knowledge*.

In other words, obtaining knowledge is presented as undesirable because it is punished by expulsion, and the acquisition of knowledge is said to be mere consumption: it is somehow already present (in fruit) and just needs to be ingested (eaten). Fortunately, religion has gradually receded over the centuries. It now plays the role of supplier of bad explanations in fewer and fewer areas: god is whatever reason cannot yet explain, which is known as the concept of "the god of the gaps." (This concept is easy to vary externally because, as an explanation, it is not adapted to the gaps it tries to fill.)

As religion began to recede, other static functions emerged around the time of the industrial revolution, which revolution helped create unprecedented wealth in Europe and the United States. Wealth can only be created by solving problems. Creativity is a wealth creation machine: knowledge creates wealth, which in turn enables the creation of additional knowledge, causing rapid progress.

This cycle is a severe threat to static functions. Is it a coincidence that new static functions started emerging soon after the industrial revolution had succeeded in creating so much wealth? Consider the following mistake of the English scholar Thomas Malthus. In his 1798 book "An Essay on the Principle of Population," Malthus noted that while an increase in a population's food supply was temporarily beneficial, it would inevitably result in an increase in population, which would, in turn, restore the original per-capita level of food production. Moreover, he posited that populations grow exponentially, whereas resources only grow linearly. He concluded that any population growth would eventually result in tragedy and starvation of the masses, as food production could not possibly keep up with the demands of a growing population.

He was wrong on both accounts: populations do not grow exponentially; especially Europe and Japan report birth rates lower than ever. Likewise, food production can increase more rapidly than linearly through *technological innovation*. Deutsch has refuted Malthusian mistakes (2012, 205-206), and I agree with him. Why is it that innovation did not occur to Malthus as a solution to the problem of feeding a growing population? On the one hand, because no-one can predict the growth of knowledge (Popper 1983, 298), and with it, innovations. But on the other hand, why do those spreading doomsday scenarios such as Malthus not realize that technological innovation *could* make a difference, as it had before, and that romantic ideas of a plentiful past were false? Because static functions prevented him from either thinking of it altogether, or seriously entertaining it as a solution. Forecasts involving the unknowable are prophecies (Deutsch 2012a, 198).

Flatheads like Malthus have come and gone since and have always made the same mistake. Paul R. Ehrlich, known for his book "The Population Bomb," is a present-day example of another victim of the same set of static functions: he predicts the same catastrophe Malthus did. His "solution" is compulsory population control should voluntary birth control and abortions not work.[38] His opinion on higher energy consumption is "[…] giving society cheap, abundant energy at this point would be the moral equivalent of giving an idiot child a machine gun." (1975) He equivocates our very means of survival with a death trap while insulting children and trying to force us into the real death trap.

Malthus, Ehrlich, et al. are not only wrong – they are *deadly* wrong. The solution is to have as many people as possible, for only people solve problems. And to help them innovate, we need as much energy as possible. Perhaps Malthus could be forgiven; population studies were a new field, and the nature of human progress was not well understood – indeed, in his time, humans had only just started making progress. But Ehrlich should have known better.

By his time, humanity had already created ingenious inventions to solve hunger problems. One such invention was the

Haber-Bosch fertilization process from the early 20th century, which still today feeds roughly half of the world's population every year. Scientists Haber and Bosch saved *billions of lives.* However, to a pessimist like Ehrlich, any such progress only delays the inevitable disaster of mass starvation. According to his logic, saving lives eventually causes the death of all, as they are crushed by the additional hungry mouths that they need to feed; the billions that Haber and Bosch saved are allegedly the billions that are going to starve. By that argument, escaping death is to cause death. Nonsense. The real cause of death is policies suggested by pessimists like Ehrlich. Haber and Bosch are heroes.

Technological innovations are often said to be "ambiguous": the same technology can be used for good or for ill. We have already encountered this misconception: it is relativism applied to technology. It grossly undervalues the enormity of the objective technological progress we have made. In reality, there is no ambiguity. A new technology, like all knowledge, either solves a problem, or it does not, and the new problems it creates, including the potential for misuse, are either better than the ones it solves, or they are not. If either of these conditions is not met, the technology is not adopted and never becomes an innovation.

Shortly after the first wave of the industrial revolution, *socialism* emerged. It is a significant enemy of wealth creation and a prime example of the denial of creativity: like inductivism, it considers all of the knowledge – in this case, wealth – to already be present. It sees wealth as a pie of fixed size rather than one that can grow indefinitely. It calls for the "fair" slicing and redistribution of the pie rather than the creation of more wealth. Such slicing and redistribution can only be realized through theft, which is a form of violence. Socialism imagines human interactions always to be zero-sum games and leaves no room for the possibility of win/win situations. Curiously, socialism associates with *environmentalism*. Many environmentalists today want to stifle economic growth and even the growth of populations because our planet allegedly has scarce resources. It has even led some to believe such nonsense as "the world would be better off without us." It is a *Malthusian mistake*, and it is another way to deny creativity: by implying that humanity should not have children, it would lead to the extinction of humanity, and thereby the extinction of creativity. Anti-*creative* sentiments are always also anti-*human*.

Environmentalism feeds off the idea of Mother Nature or Spaceship Earth (the latter term goes back to the economist Henry George[39]): the idea that the Earth is particularly suited to supporting human life and that it is human activity which destroys the "balance" of nature and makes the planet uninhabitable. In reality, the opposite is true: it is human activity, that is, the development of technology, which enables us to survive in the first place. Creativity is our lifeblood; we have no other means of survival. In his brilliant refutation of Spaceship Earth, Deutsch gives the example of Oxford, where a cold winter night could kill any human not equipped with clothes and other human inventions (2012a, 48). I agree with him that the idea that such an environment offers life support is far fetched. Deutsch concludes that the "Earth's biosphere is *incapable* of supporting human life," (2012a, 45) and that describing it as the *opposite* of a life-support system – a "death trap" – comes much closer to the truth (2012a, 48).

One could argue that winter in Oxford is particularly harsh and, therefore, not representative of the rest of the Earth. But Deutsch's argument works even for the most *pleasant* of environments. Place yourself in the most naturally pleasant environment imaginable on Earth: for example, Southern California in the spring, which features a nicely dry temperature of 21 degrees Celsius (70 degrees Fahrenheit) and virtually eternal sunshine. Only imagine it without any life support built by humans: no cities, no roads, no farms, no technology whatsoever. You have no clothes, no food, no water, no shelter. If you find a river you can drink from, you may be able to hold yourself over for a few weeks. Otherwise, you will die in a matter of days. In most environments less suited than this one, you will die in a matter of hours, even though they may be sprawling with raw materials and resources. So your survival depends not on resources alone, but *knowledge*.

Resources are plentiful, but knowledge is scarce – and what is a resource and what is not depends on knowledge. If you find yourself in an environment with plenty of resources but little to no knowledge of how to use them, you will die. Conversely, if you find yourself with limited resources but sufficient knowledge, you can survive anyway. A dry grass plain can be a death sentence or a source of heat, depending on whether you know how to make fire by rubbing sticks. An oil reserve serves no purpose without a little knowledge of thermodynamics. The sun is either a source of skin cancer or a source of energy.

But maybe nature lovers would like to try living deep in the woods so they can feel all of Mother Earth's love for them and live as nature intended? And maybe progress deniers generally would prefer to live as little as a hundred years ago, when an infected paper cut could kill you? And presumably, those who deny *moral* progress would prefer living in Nazi Germany, or at least see no significant difference to our Western society?

An editorial in the scientific journal *Nature* (2019) features a photo of British natural historian David Attenborough, "widely considered a national treasure in the UK,"[40] with the obligatory look of worry on his face. In its first line, the editorial asks, "How much do species and ecosystems contribute to the size and growth of economies?". And it says Attenborough will be the "public face" of an "expert panel" appointed by the UK Treasury to answer questions such as this one. But species and ecosystems do not contribute to the size and growth of economies. Only *people* do. And only explanations guessed by people tell us which resources, species, and ecosystems are at all relevant to economic growth. Thinking this knowledge is somehow already present in nature is *eco-induction*.

Why is it that environmentalists preach a reduction of economic activity rather than solving the problems that climate change presents? Because from the perspective of static functions, problems need to be *avoided* in order to stifle creativity. Once you acknowledge problems, you allow room for the possibility that they are soluble.

It is not a coincidence that socialism and environmentalism associate because, as I discussed, they are both static memeplexes of a similar kind. Both deny that progress and growth are desirable, with a particular focus on the creation of wealth and the consumption of resources. As a result, both are anti-human and anti-wealth and share the insidious ability to attract and compromise caring people by hiding behind a front of seemingly good causes and by appealing to good intentions and dreams of *utopia*. I guess that static functions, though generally inhibiting creativity, have nonetheless evolved to whitelist the spread of certain kinds of other static functions because their cooperation will improve their ability to spread. They can control criticism well enough to have influence over which kinds of other functions are copied successfully and which are not.

Both socialism (wealth-denial) and environmentalism (progress-denial) are powerful static functions that have been spreading for many decades. Today, after the roughly 100 million people socialism murdered in the 20th century, it is still considered by many – among them well-meaning people – a superior alternative to the free market. Some claim that socialism has never really been attempted. Right. What *has* never been attempted is a government based on the people's consent. The Westerners who fail to criticize socialism have the same moral blind spot that enables socialism in the first place.

It is difficult to estimate the kill count of environmentalism because it is ongoing, and because the destructive forces of environmentalism are not well understood, meaning when deaths do occur, people may attribute them to something else. But the case of Golden Rice is clear. Golden Rice is a genetically modified rice that differs from its parental strain in its synthesis of beta-carotene[41], which, upon consumption, an intestinal enzyme splits into two vitamin A molecules. Beta-carotene is an orange pigment that causes the rice to be orange rather than white. It is also what gives carrots and sweet potatoes their orange color.

In an article titled "How Many Lives Are Lost Due to the Precautionary Principle?", Adam Thierer (2019) presents the "Golden Rice Case Study." Scientists in Germany developed Golden Rice in the early 2000s to counter the global vitamin-A deficiency. Lack of vitamin A leads to immune deficiency, causes blindness, and eventually death, "primarily among children and pregnant women in undeveloped countries," resulting in "an estimated million deaths each year." Fortifying rice is a preferred solution over dietary supplements because rice is already a staple food in many of the affected developing countries.

107 Nobel laureates have signed a letter (Agre et al. 2016)[42] urging the environmental organization Greenpeace to drop "their campaign against 'GMOs' in general and Golden Rice in particular." In addition, Thierer quotes science writer Ed Regis saying that "one of the primary reasons it took 20 years to develop the final version of Golden Rice was 'the retarding force of government regulations on GMO crop development.'" That means due to active intervention by governments and environmentalists, 20 million lives that could have been saved, were not. That is only Golden Rice; it is safe to assume that environmentalism and regulations block countless other life-saving innovations.

If this continues, environmentalism's death toll in the 21st century will be similar to that of socialism in the 20th, if not higher. What is Greenpeace's motivation? The good intention of *ensuring that the research is safe*. They believe "GMOs should not be released into the environment since there is not an adequate scientific understanding of their impact on the environment and human health." (n.d.)

Is the failure to prevent death equivalent to causing it? Who cares? It is not necessary to pass moral judgment to realize the destructive nature of environmentalism. Rather than worrying about drawing precise moral lines in the sand, we better focus on destroying environmentalism before it destroys us. Today, the vast majority of people who desperately need Golden Rice still do not have access to it.

Thierer points out that the opposition to Golden Rice and other GMOs is a special case of a much larger misconception: the *precautionary principle*. (For a brilliant refutation of it, see also Deutsch 2012a, chapter 9.) It is a translation of the German *Vorsorgeprinzip*. German legislators invoked it in the 1970s to ban certain substances for environmental reasons. The term was later translated into English as *precautionary principle* by Konrad von Moltke in 1988 while addressing a British audience.[43] We have already encountered this principle in the context of AGI research. It states that we should "avoid everything not known to be safe." (Deutsch 2012a, 221) It is another case of good intentions killing the very people they claim to protect. However, because they are good intentions, they are difficult to criticize: anyone who dares criticize them runs the risk of appearing evil and, therefore, feels pressure to adopt them instead. This effect is at once a defensive and replication strategy static functions have evolved. Few things merit precaution as much as the precautionary principle.

What does all of this have to do with psychotherapy? Building on Deutsch's work on (un)happiness (2012a, 318-319), I suspect the primary source of mental ailments is the inhibition of one's creativity and the resulting inability to solve one's problems. Further, I guess it is the same kinds of functions that cause stagnation and death on the level of society that cause depression in the individual. That is this chapter's first central thesis.

A good way of destroying a man's creative abilities *and* denying the existence of creativity at the same time is to deprive him of his individuality. Static functions do this by shifting his focus on the collective – à la "technology should be built for the benefit of others" or "think of the environment!" But it becomes increasingly difficult to deny people's creative capacity as they make progress. In response, static functions try to *compromise*, which, as I have argued, leads to bad explanations. "Granted: science can do more than predict – it can explain – but only in terms of lower-level constituents" (i.e., reductionism). "Okay, we will not redistribute *all* wealth, but *some* of it" (i.e., less severe implementations of socialism through taxation). Or, generally speaking: "*Some* progress is possible, but it has to come to an end at *some* point."

However, there is no problem in the infinitely deep pool of soluble problems that cannot be solved by people. The only thing that can prevent the solution to a problem is a lack of knowledge. This principle is Deutsch's "Principle of Optimism": "All evils are caused by insufficient knowledge." (2012a, 212) As we will see in the next chapter, people can solve a fantastic amount of problems at once – particularly through the development of AGI. But, as Deutsch points out, the resulting state will be far from unproblematic: people will then be able to see problems which they were unable to detect before due to insufficient explanations. The more progress they make, the more problems surface, since the deeper an explanation is, the more problems it will uncover.

A static function's last hope is to convince its holder that some soluble problems are not soluble, thereby, as leading him to respect such problems too much (many environmentalists share this affliction, as do philosophers – Popper pointed out "they respect [problems] too much; they seem to believe that they are insoluble, if not taboo; and they are shocked and horrified by the claim that there is a simple, neat, and lucid, solution to any of them. If there is a solution it must be deep, they feel, or at least complicated" (2002b, 73)). Alternatively, a static function can try promising a utopia: an unproblematic state. Such a utopia is said to either be reached in the future (communism) or to be the original state from which people have diverged (Garden Eden, Spaceship Earth, et cetera). But since "problems are inevitable," (Deutsch 2012a, 64) such a state cannot exist. If the alleged utopia is ever reached (which is usually attempted by violent means), problems, including unavoidable ones, need to be avoided from then on – otherwise, it is not a utopia – which first leads to stagnation, then misery, and eventually death.

People are explainers. They naturally want to solve problems and live better lives. Problem solving and progress are inherently positive things because they help humans thrive. Though static functions should prefer[44] to keep their holders alive in order to spread, they will stop at nothing to disable their holders' critical faculties by causing them to hate humanity and progress.

Static functions infect three particular areas of human life in order to spread effectively. The first one is child-rearing. All instances of ideas in a single mind die with their holder. In other words, all functions, both dynamic and static, face the challenge of their holder's death. Some functions, on the other hand, including complex function-plexes such as languages and religions, have survived for thousands of years. All functions must solve the problem of how to survive their holder's death, and must, therefore, be adapted to turning into memes; as memes, they must then be adapted to spreading from one generation to the next (Dawkins 2016, chapter 11). They must also be well adapted to causing transmission from parents to children and between partners (Temple, personal communication with author, 2019).

Since men are a dispensable part of child-rearing after conception, I suspect that static memes single out women especially. Useful memes have an easy time spreading across generations: they solve problems, so parents will gladly share them with their children and each other. Static functions, on the other hand, must command parents to attempt to transmit these ideas to their children. From a static meme's perspective, the creativity of a child is dangerous: it is still healthy and uninfected, able to criticize with full force. It will command parents to break the child's will in order to transfer successfully to the child's mind. So children, parenthood, and romantic relationships are the first victims of static memes (Temple, personal communication with author, 2019). We should, therefore, expect irrationalities in the way we treat our children: in our relationships with them, in our schools, and in their upbringing generally. We should expect schools to drown in irrationalities and creativity-killing memes.

The second area is politics and governments. If a static function successfully hijacks a political institution, it can exert control over large amounts of people almost instantaneously: both the people within that political institution, as well as everyone governed by it. That is why, in a democracy, rather than focusing on deciding who should rule, it is more important to focus on correcting errors and removing bad actors without violence (Popper 2016).

The last main area we should expect to be infested with static functions (though I guess there are many others) is intelligence research. Understanding how creativity works involves 1) acknowledging that it exists and that knowledge creation is possible; 2) realizing that it is unbounded; 3) understanding that its constituents are conjecture and unbounded criticism. From the perspective of a static function, it is imperative to keep people in the dark about all of this.

It is no surprise then that empiricism, induction, Lamarckism, creationism, et cetera *pretend* to be solutions to the problem of knowledge creation while *denying* it by claiming that knowledge is already present somehow. (For the particular case of creationism denying knowledge creation, see also Deutsch 2012a, 104.) That is the reason why software engineering, though remarkably unaffected by static functions generally, fell victim to these misconceptions as soon as it began to attempt implementations of intelligence.[45] We should expect static functions to be *adapted* to inhibiting epistemological research of any form. They do not "want" us to figure out how creativity works. Ideally, they do not even want us to know that it even exists.

That is why companies like OpenAI do not work on AGI. Instead, it seems they are doing everything they can to *prevent* it. Their alleged goal of building "safe and beneficial AGI" (2019) is well-intentioned sounding code for the prevention of open-ended problem solving. It is a variation on the justificationist theme of attaining certainty: an impossible undertaking because problems are inevitable. They want to make sure it does exactly as it is told and does not overstep preset boundaries, which means it will be shackled and prevented from being open-ended. Static functions demand the license to enslave and murder AGI should it ever be built, all while pretending to build it: they are hiding in plain sight. If they cannot hide creativity from us, and they cannot prevent us from building it, the last thing they can try is violence: murder AGI.

Intelligence-research companies consist of physical and abstract structures built by static functions to prevent us from discovering how creativity works. Their employees are the static function's slaves who execute this immoral purpose. I say it is *immoral* because that is what static functions are: they limit error correction, which is the very definition of immoral (Deutsch 2012a, 235). This limitation is an example of the enormous causal power static functions can have. Elon Musk has said that he has invested in intelligence research not for a potential return on investment, but to "keep an eye" on the research because it is potentially dangerous (2014b).

Bostrom is another slave of static functions, which explains why his book (2017) is such a nauseatingly pessimistic attempt to snuff out AGI before it is even built. Why else does he point out that "civilian efforts in sensitive areas might be regulated or outlawed," and that "it would not be too difficult to identify most capable individuals with a serious long-standing interest in artificial general intelligence research"? He even goes on to say that "[s]uch individuals usually leave visible trails. They may have published academic papers, presented at conferences, posted on internet forums [...] They may also have had communications with other A[G]I researchers, allowing them to be identified by mapping the social graph." Why does he go to great lengths to provide a Gestapo-style manual on identifying researchers in this field if not to stop their projects? And all this even though "[h]e has long been concerned with the question: 'What should one do if one wants to have the greatest expected positive effect on the world?'" (Effective Altruism, n.d.)

Should I say despite... or *because* of this concern of his? I guess the latter, for it is a classic example of a who-should-rule question, and it begs for an authoritative answer. The goal is to *solve problems*, not do good. Should you discover a solution with reach that does great good, wonderful. But if your *goal* is to do good, not only will you fail to solve problems: you will come up with criteria you deem unchangeable, for, if they are for the greater good, changing them seems immoral. Since problems are inevitable, you thereby cement in stone a faulty criterion which, down the road, will do the opposite of doing good: it will lead to stagnation, or worse yet, decline, and, eventually, death.

Static functions love to hide behind good intentions: it makes them harder to criticize. They can exploit the decline these good intentions create because people have a harder time criticizing ideas when they go through times of turmoil and need to think about more immediate matters. Then the authors of these bad ideas can rationalize whatever terrible state they caused by claiming it was for the greater good, and anyone criticizing them runs the risk of appearing to want to make things even worse. Once we accept something as being for the greater good, we will then be tempted to demonize everyone who does not work toward it or dares question it, and we will criticize them as being blind to the allegedly manifest truth and the necessities of a moral and progressive life. But truth is not manifest. It "is hard to come by." (Popper 2002b, 502)

Darwin and Popper, on the other hand, are heroes of free thought. Against all odds, they found ways to break through static functions' dictatorship and improve our knowledge of creativity: Darwin by coming up with the first good explanation of the origin of adaptations in the biosphere; Popper by applying the concept of evolution to knowledge generally.

Physical diseases can only inhibit your brain's memory capacity or processing power; they cannot constrain your ability to think a particular thought. If they were able to do that, you would cease being a universal explainer, and with it, a person. Therefore, diet, medicine, exercise, or anything else that only has a physical effect is not going to help cure depression or other mental issues on its own.

Curing mental ailments is a software-engineering problem. That is this chapter's second central thesis. What we traditionally call a mental "disease" is not a physical disease but a rogue function or set of functions in one's mind. Many mental ailments are the result of faulty implementations that one can fix through some introspection.

Most bugs do not make a function static, and not every static function is buggy. For example, the faulty implementation of addition that has made several appearances throughout this book is wrong, but within a specific domain, it is useful and never impedes your ability to correct errors. Nor does it instruct you to avoid finding better addition algorithms; indeed, it may help you do so by showcasing its errors when used in the wrong way. Likewise, astrology is a silly mistake but does not prevent you from correcting errors. Conversely, not every static function leads to depression; for example, an avid proponent of Lamarckism or induction generally may be a happy person.

We can, for example, expect static functions to be able to trigger positive qualia: perhaps the idea of the desirability of certainty and foundations has evolved to trigger a feeling of satisfaction when finding confirming instances of a theory. Two thousand years of Aristotelian induction is a long time for a static function to evolve and improve. We can also expect static functions to have evolved another defense strategy: whenever their host is replicating another person's dynamic functions that could be threatening, his static functions trigger some dissonance in him, such as revulsion or discomfort, or instruct him to harm the dynamic function's source. They can achieve this in several ways, for example, by ridiculing or, in some cases, physically hurting the other person. Whatever the case, I suspect that when severe and sustained mental problems are present, they are the result of static functions. These functions are the enemies that lead to destructive thoughts and actions in individuals (depression) – and, when they manage to become memes, on the level of society (socialism, environmentalism, et cetera).

We do not know what the implementations of static functions are because we have only ever written dynamic ones. So how can we speed up our fight against static functions in hopes of eradicating them? Moreover, if static functions cause people to get stuck by inhibiting their creativity, how can we get them unstuck?

Consider a thought experiment involving the following two scenarios. In the first, imagine we have successfully built an AGI. Since it runs on a computer and we know how computers work, it can trivially inspect its source code and its current knowledge – the wealth of its explanations – by loading them up in a text editor. Alternatively, consider a neural interface that connects to our brains and presents the knowledge in them as functions on a computer screen, also in a text editor. For reasons I explained in chapter 7, I do not expect this ability to come out of reductionist neuroscience alone – we may have to work out our code by some indirect means – but neuroscience will play *some* role here as we need to connect the interface to our brain's hardware.[46]

By today's standards, this task presents an enormous difficulty compared to inspecting a traditional computer's memory, but we are concerned with matters of principle. For an AGI, this ability would simply fall out of its running on a traditional computer and requires no development of new technology. But in any case, the result is identical in both scenarios: a person's knowledge (AGI or human) is displayed in a text editor on a computer screen. (For privacy reasons, any person may only want to do this with his own mind, or entrust a professional with this task, for inspecting a person's memory storage without their explicit consent would be an extreme violation of their privacy. But for the sake of illustration only, we will imagine editing another person's knowledge.)

In the case of the AGI, the computer screen displaying the text editor may be part of its body or a separate piece of hardware. In contrast, in the case of the human, it could presumably only be a separate device, but such details are irrelevant for our purposes. In both cases, we have successfully loaded up the software that represents a given person, including the knowledge created at runtime, that is, during the person's lifetime. I will explore most of the advantages of this technology in the next chapter, but I want to go into one advantage as it relates to memes here. We can now inspect our functions' source code in our text editor.

Our knowledge is presumably vast and complex when displayed in this format. It may require some superficial reformatting to improve its legibility and comprehensibility to view it in a text editor, but this also does not present a problem of principle. Since we are editing a live person, his knowledge may change from under us as we edit it – but this is again no problem of principle. We can pause the function that is the person, effectively putting him in a coma with the difference of having no gap in experience; alternatively, we can instead load a snapshot into the text editor and then merge our changes into the live function. The technology to do this already exists.[47] In either case, a backup of the person should be made in case an error is introduced. (Creating backups of people is another technology whose advantages I will expand on in the next chapter.)

This technology will give us unparalleled insight into our knowledge by making it highly explicit. Today, imperfections in the structure of our functions often elude us because many of them are highly inexplicit. We cannot yet read even our most explicit knowledge. But with this innovation, all our knowledge becomes explicit, and we can examine the structure of all of our ideas line by line to look for flaws and correct them. You can think of this technology as a sort of CRISPR – a targeted gene-editing technology – for mental functions. It allows us to improve the structure of our knowledge: obvious bugs, ungainly implementations, endless loops, et cetera we can now quickly identify and fix or improve. In particular, we can identify static functions: explicitness is the enemy of static functions because it exposes them.

I do not think computer viruses are static functions because they do not infect minds. The closest we may have come to building static functions is inadvertently through poor project-management and software-maintenance skills, which can lead to massive, monolithic programs which resist change not because they are useful, but because they make it hard to criticize them. They are not *adapted* to sabotaging error correction. However, since static functions *are* adapted to sabotaging error correction, they must have specific characteristics that allow us to guess what their implementation might look like, thereby aiding us in their detection once we can read all our functions' implementations.

Since criticism is creative, sabotaging criticism can work by both sabotaging criticism directly as well as indirectly by sabotaging creativity as a whole. So although static implementations have so far eluded us, we can make informed guesses about their implementations. I guess that static functions can inhibit criticism (directly and indirectly) in three different ways and that one could defend oneself accordingly:

1. Static functions hijack the mechanism that conjectures new functions. For this case, examine the part of your creative function responsible for conjecturing new functions line by line. Static functions may, for example, append baggage to every new conjecture, or instruct the brain/computer to block or slow down for a certain amount of time (known as "sleep statements") after a new conjecture is made, or run functions causing side effects on the body which lead to unpleasant sensations (see also Deutsch 2012a, 389(. Such changes make it harder to criticize these functions later on, especially if such criticism depends on running the conjectures. Whenever you find such lines, delete them.

2. Static functions inhibit the mechanisms powering criticism in basically the same way: they slow down the program responsible for exercising criticism through baggage insertions or sleep statements. Once you have localized the program implementing criticism, look for such changes/additions, and delete them.
3. Static functions take up excessive amounts of the brain's resources (memory and processing power) in order to make it harder to allocate resources for the creative algorithm generally. Rather than conjecture new ideas or criticize existing ideas – both of which are expensive in terms of resources required – the mind is now encouraged only to run existing functions, leading to stagnation and an inability to solve new problems. This kind of implementation may be harder to identify because it requires looking through *all* code rather than just the creative program, but this is not a new problem: software engineers do this routinely using so-called activity monitors and task managers that show all currently running functions and their corresponding use of memory and processing power (though a nifty static function may evolve strategies to go undetected or undermine such an activity monitor). For example, you wake up with anxiety every morning. Upon waking, the first thing you think of is the time you walked in on your wife cheating on you with your best friend. This memory causes shortness of breath, an elevated pulse, and anxiety. You check your source code and find that indeed, there is a function that triggers every time you wake up, invokes memories of that dreadful moment, and proceeds to raise your pulse and run other functions that cause unpleasant effects, resulting in anxiety. It presumably runs right when you wake up because that is when your defenses are down, and it is a way to keep itself fresh in memory for the rest of the day. Now that you have identified the function and its location in specific lines of code, you can delete it.

Note that static functions are not the *result* of this new debugging technology – that would be shooting the messenger – but that instead, it allows you to detect, inspect, modify, and delete them. Anyone who experiences anxiety or other mental ailments must have functions causing them, for the brain is a computer, and its instructions are functions.

You may have noticed that such a technology would make it possible to extract memories (memories as in "remembered," not as in "computer memory"). This is true, and it is a premise some science-fiction narratives have explored. However, I expect that people will rarely use it for this purpose since they can instead remove associated functions that trigger unpleasant feelings. Note also that life experiences or "learned lessons" do not diminish in value: it is not necessary to endure the pain of an unpleasant memory in order to learn from it. You will know that you should not have trusted your wife, or gone cave diving without telling anyone, or any other bad memory you may have, without needing to experience negative sensations. For, any function that is that knowledge you created in response to any one of these problems can exist without the lines that trigger unpleasantness. If you deem the solution, that is, the function implementation good, you will keep it – except for those lines.

With this technology in place, software engineers can build functions that are highly adapted to finding and eliminating static functions for the routine maintenance of the mind. We may consider them *sentinel functions*. They are a special kind of dynamic function. They will help us drive static functions to extinction, prevent them from resurfacing, and stabilize our "dynamic society," as Deutsch calls it (2012a, 396). Sentinels may also find ways of turning static functions into dynamic ones, as well as improving dynamic functions through implementations that have reach and are therefore useful to have as parts of many other memes. This technology will be available to our AGI children, and to humans, if we succeed in developing the requisite brain interface. Until then, one of our best weapons against static functions is integrity; it is the only sentinel function we have developed so far.

Building on Deutsch's work, I think most (if not all) of the mental unpleasantness one can experience is caused by an inability to solve one's problems. People experience various forms of unhappiness because they are stuck in this way. On the flip-side, happiness is an indicator that one is "continually solving one's problems" (Deutsch 2012a, 318-319). But whenever one is stuck, I guess that it is primarily due to static functions. Plain old software bugs are another reason.

In summary, it becomes clear then that psychotherapy is a branch of *software engineering*. It is a software project concerning the removal of bugs and static functions.

It always has been. The reason it has not presented itself this way is that our current methods of helping someone restructure ideas in their mind and solve problems are extremely crude: rather than improving our mental code, solutions are sought by vaguely talking about problems and sometimes prescribing medicine. The idea that chemicals in the form of medicine can solve hangups or depression is a reductionist mistake. The more effective level of emergence to solve psychological problems is that of software, not hardware, and how a patient perceives the effects of medicine is subject to his interpretation. That includes the interpretation of hormonal and other chemical changes in his body.

Though mostly devoid of superstition, present-day psychology reminds one of the medical understanding of physical diseases in the middle ages in terms of crudeness. Granted, changing the code in one's mind is possible without the technology I have presented: for example, one can always conjecture new ideas that trigger changes to other ideas, resulting in code changes for the better. Likewise, it is possible through close introspection and reflection to make the implementations of some memes explicit. So in principle, psychotherapy does not depend on new technology: we can know about the implementation of our software without ever reading it directly, by making guesses about it and criticizing those guesses. We can change our software, too, via the same mechanism.

But for the moment, this is all rather crude and inexplicit. The technology I have suggested will give us the ability to improve psychological issues to the point where they are highly explicit, routine software maintenance projects. We do not yet know how ideas are stored in the brain. But we know from theory that they exist. When we do discover them in the brain, it will be a similar discovery to the DNA molecule, which was also previously known from theory (Deutsch 2012a, 373-374). We further know from theory that ideas are functions, as are all abstractions, and should expect their implementation to look as such.

Oftentimes, we do not even know why we are unhappy. Sometimes people think a move to a new city or having more money or sex will make them happy. But none of these things alone will do the job (Deutsch, personal communication with author, 2019). Happiness, Deutsch argues, cannot be induced mechanistically. It comes from unbounded creativity, which cannot be automated. Some people want to follow their gut or emotions to make decisions, which is romanticism; others want to rely only on logical, "reasoned" knowledge. But in all of those cases, Deutsch says, they judge knowledge by its type or origin and should instead evaluate it solely by its content. That requires creativity. The danger in always preferring, say, logical over emotional knowledge, is, according to Deutsch, that you may let a piece of logical knowledge arbitrarily win over a conflicting emotional piece without solving the conflict. By doing that, Deutsch points out, you are coercing yourself, which, down the road, will make you unhappy.

I believe we need explicit, software-level technology to aide creativity and eliminate static functions. With the technology I have presented, we will be able to detect the underlying problems and solve them faster and more effectively. We will be able to inspect our code and point to specific, trouble-causing lines, and either remove them or fix them, thereby curing psychological ailments. This technology is a mere possibility for humans, pending the development of the requisite brain interface. However, for AGIs, it is part of the very technology that enables them in the first place. I suspect it will play a significant role in the elimination of static functions and the sustained alleviation of suffering, and I expect our AGI children to live much happier lives than most humans do today.

Terminology

Dynamic/rational function – A function which helps a mind solve problems

Static/anti-rational function – A function which inhibits a mind's critical faculties

Function-plex – A collection of functions that help each other replicate

Malthusian/Ehrlichian error – "A population always grows more rapidly than can be supported by technology," which is a special case of the mistaken attempt to predict the growth of knowledge

Socialism – A political misconception based on theft from individuals and the control of society

Environmentalism – The prevention of economic growth under the guise of protection of the biosphere

Eco-induction – The misconception that knowledge for our survival is somehow already present in the biosphere

Utopia – An impossible, unproblematic state

Sentinel function – A function watching over the function pool and eradicating static functions

Summary

Memes are software (i.e., functions) that spreads from one mind to another.

There are two different kinds of functions: those that help their holders solve problems (dynamic/rational), and those which inhibit their holders' critical faculties (static/anti-rational).

Induction, Lamarckism, empiricism, et cetera *pretend* to explain creativity while actually denying it. Since criticism is a core part of creativity, they also deny the ability or desirability of criticism and are therefore static. People lived in static societies for most of history (Deutsch 2012a, chapter 15); outside the West, many still do.

Socialism and environmentalism are static function-plexes. Once more, the road to hell is paved with good intentions. Creativity is a faculty of individuals, not groups, which is why some static functions attempt to shift focus onto groups. Utopia cannot be achieved because "problems are inevitable." (Deutsch 2012a, 64)

Children and romantic relationships are the first victims of static functions (Temple, personal communication with author, 2019).

Static functions are one of the primary causes of unhappiness and the inhibition of progress, and they need to be eradicated. On the level of society, static functions lead to stagnation and deadly misconceptions and policies. On the level of the individual, they lead to depression and other mental ailments.

Psychotherapy is currently crude and unsuccessful because it is not considered a matter of software engineering. If we find a way to display and edit a person's functions, we should be able to cure them of mental ailments. Curing mental ailments is a software-engineering problem.

Static functions affect individual minds and, if spread far enough, entire groups, and even societies. An infected mind has a limited ability to solve problems. This limitation is what causes depression and other mental issues.

We should pursue the fight against static functions with the same urgency as the elimination of some viruses and diseases. Even after successful eradication, new static functions can appear through the variation of dynamic ones.

The importance of creativity cannot be overstated: we need it to survive. Any process that hinders creativity leads to stagnation at best, death at worst. Problem avoidance kills people sooner or later (Deutsch 2012a, chapter 17). Problem solving can save them.

Not only can our AGI children live happier lives than we do today – they may take us to the stars.

10
Living Among the Stars

> "Man will not fly for 50 years."
> – *Wilbur Wright to his brother Orville in 1901, two years before they both flew*

> "No flying machine will ever fly from New York to Paris."
> – *Orville Wright in 1908, nineteen years before Charles Lindbergh did*

When the astrophysicist Neil deGrasse Tyson spoke about the future of space colonization at a conference called "World Government Summit" in Dubai in February of 2018, he pointed out that it is challenging to fund and initiate "research with unknown returns on investments in fields not yet understood by the public." This difficulty applies to research generally, not just space travel (though I will explore the effects of AGI in this area shortly).

As I have argued, AGI research is in bad shape. Intelligence research generally is plagued with misconceptions about how humans learn that prevent progress. Narrow-AI research is the opposite of AGI research, and those claiming to work on AGI only work on narrow AI. As a result, funding is misallocated, and wealth destroyed. Some want the license to murder an AGI once built, and some of the most prominent thinkers and business executives of our time spread refuted fears about the alleged dangers of AGI.

But we know better than that. We understand the misconceptions hindering research and roughly what it would take to build AGI. We further understand that the research is safe. The remaining question is that of the return on investment. If an AGI is a person, why bother building it? We have inexplicit knowledge of how to make a universal explainer: by sexually reproducing. By "inexplicit," I do not mean "indirect," in the sense that two people have sex, and then "something happens" and voila, out comes a baby. I mean that our genes contain the knowledge of how to make a person and that we are just not *aware* of how that knowledge works: we have not made it explicit, so we have not yet put it into language. If we had, we could build AGI today.

So why not have children the old-fashioned way? Because AGI children may have it much better than human ones. With that, I turn to some more problems AGI will solve in addition to its positive effects on psychotherapy. Suffice it to say that the cost pales compared to both the advantages it will confer and the cost of other, more ambitious tasks that are funded routinely. I hope that understanding the advantages will also create a sense of urgency to pursue them. This project goes far beyond intellectual curiosity: it is a necessity.

People have explored some of these advantages, but I do not think they have understood the ramifications completely. Further, the philosophical misconceptions I have discussed have been limiting their achievability severely. This chapter is not an attempt to predict the future – instead, it is an attempt to illustrate the advantages and enormous reach of AGI as a technology.

In 2015, the Swedish video game developer Frictional Games released a game called "SOMA." Before the events in the game, the protagonist, a present-day Canadian named Simon, survived a car crash resulting in permanent brain damage. When the game starts, Simon goes to an appointment for a preliminary brain scan at a laboratory in hopes it may help cure his affliction. He sits down in a chair, and an apparatus is lowered over his face. They reassure him: the scan should feel about as uncomfortable "as getting your picture taken."

The entire scan takes only a few seconds. As it is administered, Simon sees a bright flash. Once the apparatus lifts to clear his face and eyes, he suddenly finds himself in a completely different room, with the doctor gone. Playing the game as Simon, you come across a woman named Catherine who communicates with you through a computer. It is now the year 2104; the Earth was recently struck by an asteroid, killing all life on the surface and destroying the atmosphere. You are among the few survivors in a research facility on the bottom of the Atlantic ocean, all of whom will eventually run out of resources and die.

Catherine is a researcher at this facility. You learn early on in the game that she does not have a human body, but can instead inhabit various robots and computers by uploading herself to them. She communicates with you through a computer, not as a means of remote communication, but because *she is running on that computer*. Simon is taken aback by this, but Catherine points out that he is not exactly human, either. Simon's brain scan was used by a rogue computer program to bring him back to life. His mind inhabits a new body that is part machine, part human, but his experience from the brain scan to this moment was uninterrupted.

As the game progresses, Catherine enlists your help in finishing her life's work: the launching of the so-called "ARK," a computer stored within a capsule that can be launched into space and then powered by solar panels for thousands of years. It is humanity's last chance of survival. She is also the architect of the virtual world running inside the ARK and prepared brain scans of her colleagues who agreed to live in this world. You and Catherine set out to find the ARK and launch it into space in order to save humanity from extinction.

Though the game takes place in a gruesome, dangerous world, it is beautiful. It masterfully explores what it means to be human. And it is optimistic: it shows that even in the darkest of times, people can overcome tough challenges by creating knowledge. Remarkably, the developers realized that a person is a running program, independent of any particular hardware. Some solutions to the problems you encounter in the game involve taking another brain scan and uploading yourself to a new body while retaining your identity the entire time.

This technology is whole brain emulation. Let us not dismiss this idea because "it's just a game": that would be judging an idea by its origin. As I said, the ability to create snapshots of a person will automatically be available to our AGI children, because we already know how to create snapshots of running programs on computers by backing them up. The same technology may be available to humans if we develop the requisite brain interface. Once a human mind has been transferred onto a computer, it is qualitatively indistinguishable from an AGI. For this reason, I will not distinguish much between humans and AGIs for the rest of the chapter. That is because both are people. It does not matter if a person ends up on a computer through a transfer from a human body, or if he was born on that computer by starting the creative program on it. I shall refer to the ability to transfer between bodies as "being *untethered*," while I will call the present-day human condition of being unable to do so as "being *tethered*." For now, we humans find ourselves tethered to our bodies, whereas AGIs will be untethered by default.

Imagine yourself living on a computer; that is, replacing your human body with that of computer hardware. You will not be satisfied with *any* computer for a body. For example, if you find yourself on a stationary desktop computer, you will sooner or later find this condition insufficient, perhaps claustrophobic. You will want a computer with input and output devices – at least cameras for eyes, microphones for ears, and speakers to communicate with your surroundings, as well as limbs that enable you to walk around and manipulate your environment. That is where robotics research comes in: its primary role is to make living on a machine comfortable, desirable, and superior to living in a human body.

Though the human body is a decent machine in terms of navigating and manipulating the environment, it is a mistake to think it is *adapted* to this purpose. It is only adapted to spreading its genes through the population. It is not adapted to providing comfort, efficiency, reliability, health, longevity, or any other such desirable quality either. It quickly gets sick, infected, damaged. To the degree any desirable qualities have been conferred onto human bodies, it is only because they help genes spread through the population.

I have discussed Dawkins' view that organisms are the slaves genes use to spread through the population (Dawkins 2016, chapter 2). That is the purpose of human bodies as well. As Deutsch points out, a slaveholder needs to keep his slaves alive so that they continue to work for him, not for their benefit

(2012a, 92). I should add that he improves their wellbeing only to the point where they can perform the work to an acceptable degree. This degree can be much lower than what promotes comfort or health, which can result in tremendous suffering.

Genes jump ship before their body dies by commanding it to have offspring. This way, Dawkins argues, some genes spread for thousands or even millions of years, while their organisms, such as humans, are but temporary and disposable instruments genes use to live as long as they do. (Dawkins concludes this is the reason selection in biological evolution cannot happen on the level of the individual organism: the organism is not the replicator, the genes are. The entire organism is too large and short-lived to qualify as a unit for selection.)

In addition, Dawkins has pointed out that there is no selection pressure to ensure the human body's survival or health beyond the age of reproduction. Therefore, once we know how to untether ourselves, I suggest dropping human bodies in favor of artificial bodies entirely. Artificial bodies are likewise disposable – but it is *we* who own and control them, not genes.

As I discussed in chapter 5, it is ideas, not genes, that determine our behavior. The evolution of ideas deserves careful study. Dawkins (2016, chapter 4) has likened the gene building a body to a programmer running a program on a computer: once the program runs, says Dawkins, "the computer is on its own," and the programmer has little influence over it during runtime. After the genes' precise control of embryonic growth, their influence fades, "and [they] can only passively sit inside" the body.

What does this mean for the nature/nurture debate in psychology? I think it means both sides are wrong because they detract from the real origin of our knowledge: ourself. People's "nature" in the form of genetic knowledge – and with that, genetically determined behavior – plays a small role compared to the knowledge they create at runtime. And "nurture" in the sense of knowledge created through influences from the environment is induction again: there can be no knowledge streaming into us from the outside. We are its creators, not its passive recipients.

In any case, there are many more advantages to being untethered than meets the eye. The first thing to note is that *it immediately cures all physical ailments*. It is superior in this regard to other technologies envisioned to enable the transfer of a mind onto another *human* body. Why transfer to a human body when you can transfer to an artificial robot body? I guess this idea rests on the notion that people are hardware or human bodies specifically, and that, therefore, continuing to live would presumably only be possible or at least desirable in a human body. But that is not so. It is not clear where the human target body would originate; presumably, it would need to be constructed. Constructing a human body is much more complicated than building a robot because we understand how to use metal and silicon better than bones and tissue.

Even if building a human body from scratch presented no challenge whatsoever, it is not *desirable* to live in a human body. It is too vulnerable: it can get cancer, age, break down, get infected and sick, go blind and deaf, is the target of a myriad of hostile bacteria and viruses… There are no bacteria or viruses adapted to infecting inorganic hardware such as robots. Again, genes have little incentive to keep a human body healthy beyond the age of reproduction. The whole point is to escape this prison that our genes have kept us in.

Having a machine body immediately confers all the benefits of technologies that apply to machines because we know more about them and how to maintain them than we know how to do with human bodies. So, instead of placing yourself in an ordinary desktop computer, imagine yourself on a computer that has many more features: for example, a robot with eyes, ears, speakers, and limbs. When I say "robot," I mean a machine that includes both a computer and additional hardware used to move and interact with the environment. It is not programmed to do anything on its own. The human body may inspire its design if you so choose, but you will soon want to make improvements. For example, rather than storing the processor and memory of the computer in the robot's (your) head, you may choose to place them in safer locations in the body: perhaps in the torso behind a thick metal shield for protection. You may also choose not to place them next to each other for safety reasons, should an accident damage a whole region of your torso. In the event, you are free to design your body in any way you like.

Note that you are not commanding a robot in the conventional sense: you are not remotely controlling a lifeless body. You *are inside* that body. It is yours. You command it the same way you used to command your human body, and your perspective will continue to be that of a first person. When I say "you," I mean the creative program that is you, not your body – robot or human.

Learning how to use your new body is not revolutionary. After all, you learned how to use your human body when you were a baby. As you try out your new hardware, you will find that you never feel hungry, thirsty, or even tired. There is no need to buy, prepare, or eat food. The only thing required for sustenance is electricity, which is cheap, and can be acquired through solar panels on your body, with an option to plug yourself into an outlet should you need to. Though you may continue to enjoy having a home, you do not *require* shelter anymore to survive. You cannot feel hot or cold. You will not sweat or shiver. You do not require sleep, which gives you back a third of your day. Picture the cumulative advantages of this fact alone on humanity in terms of how much time people can spend on problem solving instead of sleeping. Sleep places an upper limit on how many problems you can solve every day; as it stands, sleep must eventually interrupt any creative process. But in a new body that does not require sleep, you can theoretically make infinite progress at a time. Should you like to rest for whatever reason, you still can: it will be easy to turn off your sense devices for a while to enjoy the calm presence of your mind without any disturbances from the outside. I guess fatigue is an artifact of *human* bodies, so there will not be a need for rest anymore.

Your new body is not impenetrable – it will have flaws, but rather than being stuck with those flaws as you would with a human body, you can address them. Accidents still happen, and parts of your body may break at times. If you fall and break your robot arm, it's not a big deal: it does not hurt, and you can replace it. A replacement will be much cheaper than having surgery, and recovery is almost instantaneous: it is a *hardware-maintenance project.* It is as simple as going to the store and buying a replacement for your arm.

Pretty soon, you realize that basing your body's design off the human body is not a feature, but a *limitation.* You will not miss or want your human body anymore, neither whole nor any part of it. Your new body is modular: you can attach new limbs and devices and learn how to use them. We do not yet know how to make human bodies fly, but we do know how to make machines fly. There will be no need for cars, or even airplanes for that matter: your body will be capable of all of these things (though shortly we will learn that there will be no need to transport people on airplanes anymore for another reason).

I suspect that society changes to some degree once people shed their human bodies. Since food and water are now useless, these industries are not needed anymore. Health insurance is optional; it is rather like some people get insurance for their laptops. Most people do not do it because it is not prohibitively expensive just to replace them, and it is not that big a deal when they break. Health insurance today, on the other hand, can be prohibitively expensive.

Other industries will spring up instead: that of ever more sophisticated and nimble hardware that people can buy in order to upgrade their bodies or even replace them with entirely new, better bodies. People are not limited to the size of a human anymore. You can explore the world in everything from extra large to extra small bodies, depending on what you want to do. If you want to climb Mt Everest, a bigger, sturdier body may be beneficial; in any case, you will not need to worry about getting enough oxygen at such high altitudes. If you want to explore an underwater cave's nooks and crannies, you may choose a body the size and shape of a fish specifically designed to withstand high pressure. And with this body, there would be no need to worry about breathing under water.

I suspect that learning how to use new hardware will be akin to learning how to play video games, particularly ones from a first-person perspective: it takes a little bit to get the hang of it, but you get a decent grasp of the controls soon enough. Moving body parts involves running functions that have side effects on that body, and a creative mind can conjecture these functions. That is how humans use their bodies.

Being untethered, you may find yourself browsing a hardware store with all kinds of bodies of various functionalities and sizes. As you browse the store, you find a model you like. According to its specifications, it can fly twice as fast as your current body. You tell the rep you would like to try it out. "Certainly," he says, "I can help you with that. Follow me." He unpacks one of the models you are interested in and pulls out a cable. You plug it into an outlet in yourself and connect it to the model. The program that is you is paused and then copied onto the new robot, where it resumes. Presumably, this will not take very long, but it does not matter either way: to you, the transfer will feel instantaneous because you are not running during the transfer.

You awaken in the new body just a couple of feet away while your old copy is dormant. "Go ahead, try it out," the rep tells you. "We're open for another hour." This body has additional limbs and features you can control. You learn how to use them pretty quickly. You exit the store and try things out. It takes you about a minute to fly to the next town over, roughly 50 miles away, which is indeed twice as fast as your current body.

You return to the store with a big smile on your face – or whichever way people will communicate enthusiasm then. The rep sees your reaction. "Should I ring you up? You can trade in your old body so we can reuse the hardware. I can probably get you a discount in exchange." You agree. You already have a collection of bodies for various purposes at home, and this one is better than your previous one. The rep erases the old copy of you from your old body, and you leave the store. If you had decided not to buy the new body, your old copy would have been replaced by a re-transfer of your current self onto your previous body. The copy was only kept as a backup in case something went wrong while trying out the new hardware. If, say, you had had an accident that had resulted in death, the store rep could have trivially revived you in your old body since it was still at the store.

As I have said, we have the universal ability to create backups of any computer program at present. This functionality does not require any new technology. Combined with being untethered, *it effectively makes you immortal.* When I say "immortal," I do not mean you will be cursed to live forever. Death is still possible and synonymous with forever terminating the program that is you, for example, through a fatal accident damaging your hardware. The historian and philosopher Yuval Noah Harari (2015) has called this condition "amortality." I prefer the term "effective immortality," but I will refer to it as "immortality" for brevity. Remember, you are a program (Deutsch 2016b), so you can make a backup of yourself. The technology to do this will be available to you in the same way you can create a backup of your desktop computer today. Every backup will be a dormant copy of yourself, though it will not be *you*: you are a *running* program, so, as long as the copies are not run, they are but blueprints which have the potential to instantiate you. They are potential "yous."

Whenever the body you run on breaks down to the point you are no longer running, that constitutes an emergency. That body either needs to be repaired at least to the degree that you can run again (i.e., regain consciousness), or your most recent backup needs to be installed and run on another body. There should be automated checks in place to assist people in such an emergency, and automated hooks to reinstall a backup.

Unfortunately, the quest for immortality has encountered resistance as anti-aging research has progressed over recent years. This resistance has been caused both by static memes wanting to put a cap on every individual's creativity, as well as the *appeal to nature*, a special case of the misconception of judging an idea by its origin: "death is good because death is natural." Just because something originates in nature does not mean it is good. Cancer can have natural causes, such as UV light, and unnatural causes, such as smoking. It would be wrong to conclude that cancer caused by the former was good, and only cancer caused by the latter was bad. So natural or artificial origins cannot have any bearing on whether something is desirable.

Not only is death uncontroversial, but it is even said to give meaning to life. If life was not short, the argument goes, there would be no value in it. But that is not true either: most things people do and value rarely concern the prospect of death even as an afterthought. In one's youth, by many judged to be the most meaningful time of their lives, death seems a million years away. Meaning is *created* by *solving problems*, not derived from scarcity. If people derived meaning from scarcity, they would want to die *sooner* in order for their lives to become *more* meaningful.

People may think they would not want to live past a certain age because they would have done everything there is to do. However, curiously, the stated timeline often coincides with the age of a typical death, usually between 80 and 100, thereby contradicting the argument that life is only meaningful because it is short. And the idea of running out of things to do is false: it rests on the misconception that an individual's progress must come to an end. In reality, progress can be unbounded for individuals. Following Deutsch (2012a), I believe there is always an infinity of unsolved problems that concerns you personally, and there will always be an infinity of ever more meaningful, more awesome problems and solutions you can discover, leading to an ever-improving life.

The financial implications alone are staggering. Some long-term investments can realistically double approximately once every ten years. If you invest \$10,000 as a *mortal* 20-year-old, it will have grown to roughly \$100,000 by the time you are 65 and want to retire (accounting for 3% annual inflation and 15% tax). Then you need to start cutting into that reserve in order to live, and you will not be able to enjoy any growth in your financials beyond your death. But as an immortal, the growth of your investment would never have to come to an end. If you invest the same amount as an *immortal* 20-year-old, the same investment can grow to over \$3,000,000 by the time you are 125. Just ten years later, it will be roughly \$6,000,000. Of course, this is "only" the growth of wealth; you will also accumulate intellectual wealth, which can compound over time as well. On a personal level, immortality is a necessary condition for what

Deutsch (2012a) calls a "beginning of infinity," and it is quickly and cheaply achieved through backups.

I guess the main reason some people want to die eventually is the understandable dread of being old and sick forever. The prospect of sickness "until the final flicker of life's ember," as the old Sinatra song goes, to live out one's final years with one's thoughts dominated by medical concerns is horrible. (Philip Roth's *Everyman* describes aging beautifully.) Aging, sickness, and mortality are tragedies. They are the only tragedies that have so far spared *no-one*. They have caused more deaths than all genocides that have ever been committed. They are the single biggest genocide – evil incarnate. Hardly anyone feels any urgency, despite roughly 100,000 people dying each day of "age-related causes."[48]

Consider the great innovators and thinkers death robbed from us. The contributions they could have made beyond their death – in collaboration, even: Darwin working with Dawkins and Popper on improving the theory of evolution, or Newton working with Einstein on unified field theory. The knowledge and wealth any one person could accumulate forever! The joy of living a healthy and prosperous life without bounds!

Today, we look back on human history and shudder in horror of how violent it is. Because of the progress we have made, we sometimes forget how violent it still is today – almost entirely due to biology. Infinite progress for the individual is contingent upon immortality, and the solution to death, disease, and aging is not resignation and acceptance, but untethering. Pain and suffering are not things to be accepted but fought and extinguished. And progress really is happening in almost all areas of life – at an accelerating rate. If you picture a person from 1000 years ago marveling at our lifestyle today, you get a feeling for what it might be like to live even 200 years from now.

Most anti-aging researchers avoid the topic of immortality because it is too science-fiction-y. I guess this is because it is difficult to imagine keeping *human bodies* alive forever (as opposed to software). As a result, it is more respectable to talk about life *extension* instead. These efforts happen on a *biological* basis: the goal is to extend the life of human bodies. Achieving this goal is better than nothing because every decade a person gains gives him more time to solve death. But the long term goal must be to get rid of the cages that are our bodies. Nobody seems to notice or care about the repetitive nature of the tasks our bodies impose on us: you have to sleep at least seven hours a day, eat and drink several times a day, and you have to maintain your body by exercising it regularly, or else it will age even faster. Your human body is not you – it is an appendage that you are stuck with, that forces you to take care of it, and eventually pulls you with it into death. If you think death should be avoided, you will realize the urgency with which we should pursue this project. We need to build bodies that help *us survive* rather than our genes spread.

I guess that my proposal for anti-aging in the form of immortality through backups and living on machines will generally be rated high in terms of desirability, but low in terms of feasibility.

Feasibility seems low for *humans* because we know next to nothing about how to build the requisite brain interface to initiate the life-saving transfer of the mind and untether ourselves. But we know more about how to build an AGI, both in comparison to building the brain interface and stopping or at least prolonging biological aging, so the feasibility for our AGI children should be ranked higher compared to both.

In terms of desirability, it should rank higher than current anti-aging research: as I have argued, human bodies are limiting, uncomfortable, disease-prone, and much harder to fix and maintain than machines. They are a genetic artifact, and we would be better off without them. Our knowledge of machines is far superior to that of human bodies. In short, I suggest continuing present-day anti-aging research for humans should efforts to construct a brain interface fail. However, our main priority should be creating AGI so our children can be immortal and persistently healthy. If we are lucky, there may even be a chance our AGI children will be able to help untether us humans somehow.

There are initiatives to back up minds by reconstructing brains on a molecular level. Let us ignore how hopelessly reductionist this seems and concede for the moment that it could, in principle, be done. We shall nonetheless prefer to replace brains entirely. Impressive they may be, they rot, they are vulnerable, and eventually decay. We already know the brain is a computer. We have other computers we understand much better, both in terms of functionality as well as maintenance. Let us use those instead. It is impossible to maintain any one body forever, biological or not, and the transfer of software between computers allows for cheap and fast replacement of bodies.

Immortality will surely raise concerns about overpopulation, particularly on the part of environmentalists. Such concerns are invalid for the same reason all previous concerns about overpopulation have been invalid: they rest on the Malthusian mistake of prophesying or discounting the future growth of knowledge (Deutsch 2012a, chapter 9), thereby portraying us as depending on available resources instead of the knowledge we create. It is unclear if, once untethered, people will choose to have the same amount of children as we do today, but let us assume for the sake of argument they will. Even while *tethered* to human bodies, people could keep multiplying virtually ad infinitum *on Earth*, as long as their knowledge of how to use Earth's resources improves. As I have said, resources are plentiful; it is the knowledge of how to use them that is scarce.

But this criticism still lets environmentalists off far too lightly because they reduce people to physical bodies moving through and interacting with the physics of Earth. In reality, people are *programs*: they are abstract, not physical. Once untethered, there is no strict requirement to inhabit any particular body, or set of bodies, or to consume resources other than electricity, or to even stay on planet Earth. So of all people, mistaken as they are, environmentalists should be the *most* excited about immortality. The decoupling of people from human bodies is the first step in their *independence from as much of nature as possible* and helping people do this should be environmentalists' top priority.

How often might you choose to back yourself up? It depends on your attitude toward risk. You can do it once a day if you like. I guess there will be companies offering services to automate and store your backups. There is no reason your body could not be connected to the internet directly, and an encrypted snapshot of you could be uploaded to a server once a day. If something happens to you that destroys you along with your body, your last backup can be copied onto a new body and run, thereby bringing you back to life. You may lose up to one day of experience, but that is a small price to pay for immortality. If you find a sudden jump in time or location within your uninterrupted experience, you may have been restarted from a backup. The first thing you may want to do in such a scenario is to find out what happened to your previous self.

This scenario brings up the related question of duplication. It is, again, trivial using present-day technology to create a copy of any program running on a computer, including you (if untethered). The thought experiment of you trying out a new body at a store depended on this ability, but there the technology is used in a way that does not result in two instances of you running at the same time. Deutsch doubts that copying oneself would be done casually, because every time you decide to copy yourself, you cut your net worth in half (2018a, 11:00 – 12:00). I do not recommend doing it lightly, either: there would be someone else who knows everything about you, including your secrets and fears. That someone, Deutsch argues, starts as you but then diverges and becomes a different person over time – just like you are only a faint replica of who you were, say, 20 years ago. So, you would need a good reason to copy yourself. Deutsch concludes there may be certain edge cases where copying oneself is beneficial, but that people will not do it often.

Since your body is connected to the internet, you can communicate with others instantaneously without the need for any external device. It is only a matter of software. When such communication software is installed on your body's computer, for example, through a download you initiate from the internet, you can then learn how to use it and communicate in this way. It is effectively mind-to-mind communication; only there is nothing mystical or spiritual about it. Since software is universal, there is no limit to the kinds of software that can be installed within you in this way. Entire worlds of virtual reality will be available to you for your enjoyment. Rather than putting on a virtual reality headset, you will find yourself *inside* the virtual reality whenever you choose to run such a simulation. The experience can range from what can adequately be considered television that you watch in your mind, and that does not require an external device to immersive, interactive experiences in artificial universes you or others design.

I guess the main appeal of virtual reality will be to do things that cannot easily be done physically using the current level of technology at any given time. Some people may be born into such worlds without any physical body and may never desire to leave them. You can choose to spend large parts of your life in such virtual realities, or you may choose to explore the real world instead.

These two options are not as different in principle as they may seem: recall that, while there is an objective reality, you cannot perceive it directly. You can only perceive it indirectly through your conjectured theories of how reality works, resulting in a virtual-reality rendering of the world around you that resembles parts of reality with reasonably high but imperfect accuracy. (This is no different for human bodies; it is this virtual-reality rendering that you see in front of you whenever you open your eyes.) Your immediate surroundings, both in the real world and a virtual world, are accessible to you only through your interpretations of sense data (see also Popper 1983, 146).

Therefore, it makes no practical difference when a virtual-reality rendering of the objective reality is replaced with another virtual reality rendering of an artificial reality, as all that is required is a different source of sense data. You can always focus your attention back onto the real world by instructing a program to replace virtual sense data with real sense data. We already do this routinely, albeit crudely, when we imagine the scenes from an immersive book, and when we dream, and less crudely when we watch television or play video games.

Though the untethered life will be revolutionary, I suspect that on the level of society, some things will nonetheless remain the same. Money will still exist in the real world and will be introduced into virtual worlds. Companies will likewise continue to exist. There will be software companies working on virtual worlds and software patches that you can download for yourself, as well as hardware companies working on bodies for people to inhabit. Governments will still exist. The role of borders may change, however, for reasons I shall come to now.

In order to understand your new life better, it helps to repeatedly recall that you are a running program, that is, software. There is a plethora of present-day technologies we have already built around software. Sending a snapshot or backup of a program from one device to the next is just one of them. That is how you tried out your new body at the store. There is no theoretical limit to how far away that new body could have been. You can send a snapshot of yourself anywhere in the world in the same way, as long as there is a target body to which you can upload yourself.

How might this work? If you would like to go to Tokyo for a day, you can instruct a program to create a snapshot of you, pause yourself, then send the snapshot over the internet to a robot body available for rental in Tokyo, and then run your program once it arrives. *Travel is effectively instantaneous* and costs only the price of renting a robot for the duration of your trip. I suspect that all mid to long-distance travel will be done this way. I further suspect that this will have significant implications for borders between countries because unless a country cuts itself off from the internet (which would be economic suicide), it will have a much harder time enforcing borders than it does today.

Hardware will be designed to supply bodies for all kinds of environments: you can explore the deep sea, the Earth's poles, and even its atmosphere. Before you explore somewhere secluded or dangerous, you may choose to create another backup of yourself, although this is no concern if your backups are on a schedule. If your body allows you to fly, you could indeed explore the atmosphere. If it is made to withstand a wide range of temperatures and low pressures, you can even exit the Earth's atmosphere and explore space. Remember, you do not require oxygen or food. You can fly to the moon if you like, though this may take a while at such long distances. It took you a minute to fly 50 miles to the next town over. That is 3000 miles per hour, which is pretty fast. But the moon is just under 240,000 miles from the Earth: at the same speed, it would take you 80 hours, or 3.3 days, to get there.

You have three options to tackle the boredom you might experience on such a long trip to the moon. All of them take advantage of the fact that navigation does not require creativity and can be outsourced to a program in order to automate the trip itself.

The first is to "turn inward" (not in a spiritual sense): as you travel through space, you can run virtual realities and immerse yourself in them.

The second is to instruct a program to pause you, thereby rendering you unconscious as you travel without any sense of the passage of time, then wake you up a few minutes before you arrive. There will be no gap in your experience: from your perspective, you will appear to have instantaneously jumped from just outside the Earth's atmosphere to just a few hundred miles above the surface of the moon.

But a third option is available to you. It is difficult to exaggerate its implications for space travel and exploration. Instruct a second body (not yet containing a person) to navigate to the moon independently. Do your earthly business or whatever you please in the meantime. Once it arrives, establish a radio signal with it, and upload yourself to it. This procedure follows the same logic as your instantaneous trip to Tokyo; only this time, it happens not over the internet, but over radio waves traveling through space at the speed of light. It would likewise feel uninterrupted since you would not be conscious during transmission.

Once again, we find that no new technology needs to be developed for this: today, we already routinely communicate with our satellites and rovers on other planets via radio waves, upload programs and patches to them, and download breathtaking photos they take. We simply use this technology to upload a program: you. Once the second body arrives on the moon and you establish a radio signal, it takes you all of 1.3 seconds to get there. There is no need for any additional physical transportation to get back to the Earth. You upload yourself to the body that remained there, leaving your current body on the moon. This way, you can always go back and forth in a matter of just over a second. People *can* travel at the speed of light.

Step 1: A piece of hardware is transported from the Earth to another celestial body (not to scale).

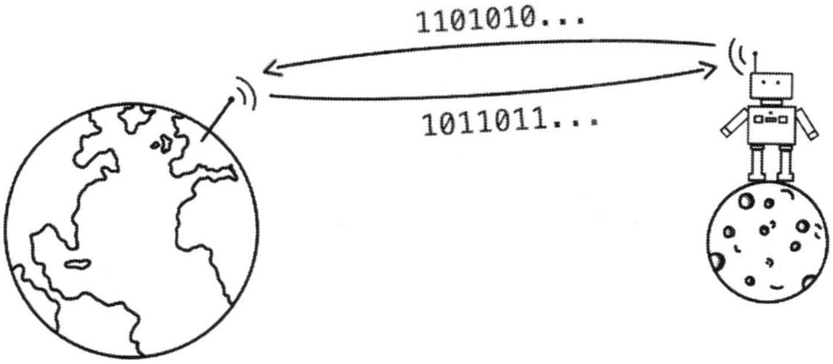

Step 2: Once a hardware presence is established, a person, depicted here as binary code, can upload and download himself to and from it, thereby traveling through space at the speed of light.

I shall call this ability of software, including people, to travel at the speed of light *communications travel.*

Most think that people cannot and could never travel at the speed of light, because, as Albert Einstein's special theory of relativity explains, when an object *approaches* the speed of light, its mass increases steeply. *At* the speed of light, its mass would become infinite. Since it would take an infinite amount of energy to move an infinite amount of mass, physical objects cannot move through space at the speed of light. Why does this limitation not apply to people? Because they are not physical objects. They are abstractions that happen to *inhabit* some physical object. Their mass does not increase as they travel through space, because they have no mass. And therefore, we do not run into the problem of having to move infinite amounts of mass to reach the speed of light. People can travel anywhere at the speed of light, like software routinely does through radio waves, as long as they have established radio communication with the destination. Thinking that people cannot travel at the speed of light rests on the reductionist misconception that people are physical objects.

Of course, you are not limited to the Earth's moon. Other planets in the solar system and their moons are now accessible for exploration. Communications travel reduces space travel time to the time it takes to physically move the desired body anywhere *once*, plus the time it takes light to travel between destinations whenever it is possible to establish a sufficiently strong radio signal.

We presently have four rovers on Mars: Sojourner, Opportunity, Spirit, and Curiosity. As of August 2019, Curiosity is the only rover still active, while the other three completed their missions before losing contact. During a 2018 dust storm, Opportunity went into hibernation mode, from which it never emerged. Curiosity's mission is to search for evidence of life, survey Mars' climate and geology, and study planetary habitability for future exploration by humans.

Rovers have several advantages over stationary landers in that they can examine more territory and be directed toward interesting features. During the winter, they can navigate to sunny positions to expose their solar panels and replenish their energy. They have sent stunning photographs of Mars' surface back to Earth.

On the other hand, rovers are slow to move, but make up for their lack in speed with their ability to climb rocks and traverse a rugged landscape. Further, rovers operate slowly due to communication delays. While communication with machines on the moon is almost instantaneous, Mars is too far away for that:

depending on its position relative to Earth in their respective orbits around the sun, a message can take anywhere from 3 to 22 minutes to travel from one planet to the other.

Despite state-of-the-art algorithms installed on Curiosity, it has limited autonomy: humans have to check up on constantly it to ensure it is not stuck or doing something that could jeopardize the mission, in which case they can remotely instruct it to change course. It will often wait for instructions. That time adds up, given the latency in communication. As a result, science progresses slowly on Mars. Without the ability to create knowledge at runtime, rovers will always leave something to be desired in terms of abilities and speed generally. In other words, we need people on Mars.

Space is seeing a resurgence these days. NASA has started an initiative to send people to the moon again in order to "commercializ[e] low-Earth orbit" and lay "the foundation that will eventually enable human exploration of Mars." (NASA, n.d.-b) In 2018, the US Pentagon was directed to begin plans for a Space Force: a new, independent military service branch for operation in space. NASA intends to send another rover to Mars in 2020. It is based on Curiosity's configuration and will collect samples from Martian rocks and soil, which it can then store in a "cache" on Mars, from where they can be returned to Earth as part of a future mission to look for signs of ancient life. The rover contains computers to process information, has internal heaters and layers of insulation for temperature control, cameras for eyes, and antennas for communication with Earth. The launch is scheduled for July or August of 2020 because the constellation of the Earth and Mars will make for a shorter distance.

Public relations included a "boarding pass to Mars," for which I signed up online along with thousands of others. Our names will be etched onto a microchip, each line of text thinner than a human hair. NASA can fit about one million names onto the chip. Nobody will ever be able to read the names with the naked eye, but this is not about that. Our names are going to make it to Mars when the rover lands in February of 2021. At the present moment, that is as close as it gets to going to Mars in person.

An example of a "boarding pass" to go to Mars as part of the 2020 rover mission

That may change soon. In 2016, the American aerospace manufacture and space-transportation services company SpaceX, founded by Elon Musk, announced its plan to begin the colonization of Mars. Their goal is to send people to Mars as soon as 2024.[49] In addition, NASA "is developing the capabilities needed to send humans to an asteroid by 2025 and Mars in the 2030s." (NASA 2014) In the not too distant future, we could build settlements there.

Since Mars has a thin atmosphere compared to the Earth's, any settlements built on Mars will need to be shielded against the deadly cold surface temperatures, intense radiation, and lack of atmospheric pressure. Long term scenarios see Mars as a potential second home for humanity. Such a second home is one answer to the problem of the survival of our species: we currently have all our eggs in the one basket that is the Earth. If a sufficiently large asteroid strikes the Earth, it could lead to our extinction.

I have discussed Deutsch's argument that a human unaided by technology will die on Earth in a matter of days, if not hours (Deutsch 2012a, 48). Therefore, it is a mistake to think that Earth is a particularly suitable and friendly environment for humans. It is, however, much friendlier than Mars, since the human body is adapted to it and not to Mars. For one thing,

Mars' atmosphere is about 100 times thinner than Earth's. If you stepped onto Mars without a spacesuit, your blood would begin to boil in a matter of seconds due to the lack of atmospheric pressure (Abram 2019). There are other complicating factors, such as lack of oxygen and the toxic, finely grained, abrasive dust covering the surface, which, when inhaled, slowly cuts into your lungs. Should you survive that, the razor-thin atmosphere cannot shield you from cosmic rays and the sun's deadly radiation and Mars' freezing temperatures will eventually kill you (Abram 2019).

At some point, if settlers and their descendants want to thrive on Mars, they will want to "shed their spacesuits" (PBS Space Time 2019) and roam free. Among other things, they will need water for sustenance. A proposed solution to this problem is the concept of *terraforming*: the process of gradually turning Mars from an extremely hostile environment into an Earth-like environment. Terraforming is said to be primarily a matter of recreating similar atmospheric conditions. There *is* water on Mars, but due to low temperatures, it is frozen. If you heat it, it will not turn into a liquid but will "sublimate directly from ice to gas" (PBS Space Time 2019, 2:10) due to the lack of atmospheric pressure. But with an atmosphere similar to the Earth's, liquid water could exist, radiation deflected, and greenhouse effects triggered to raise temperatures.

How might one thicken Mars' atmosphere? It used to be much thicker, so it might be possible somehow (PBS Space Time 2019). One theory explains its all-but-disappearance through its materials collapsing into non-atmospheric reservoirs such as polar ice caps, which, unlike Earth's ice caps, contain carbon dioxide ice in addition to water ice. If this carbon dioxide could somehow be released back into the atmosphere, e.g., by triggering a nuclear explosion at the polar ice caps, it could then increase the pressure and temperature, thereby making the planet more habitable and possibly allowing humans to survive on the surface (PBS Space Time 2019).

Unfortunately, recent results from the Mars Reconnaissance Orbiter and the Mars Odyssey Spacecraft have been interpreted as suggesting that most of Mars' atmosphere did not collapse into surface material, but was instead lost to space (Jakosky and Edwards 2018). As a result, there may not be enough carbon dioxide in surface reservoirs to terraform the planet. Furthermore, carbon dioxide stored in polar ice caps is buried too deep to be released through nuclear explosions alone, and even if it could be released that way, the amount stored today is still too low by a factor of about 100. Other carbon-dioxide reservoirs on the remaining surface amount to less than that.

Of course, even if there was enough carbon dioxide, humans could not survive on it alone: they need oxygen, so further changes to the atmosphere would be required. That is not to mention that owing to Mars' comparatively low mass, its low gravity would lead to decreased bone density and muscle atrophy. And even if we do find a way to address all of these issues somehow, successfully terraforming Mars to the point of habitability without protection may take thousands of years (Abram 2019, 12:38).

These difficulties have led some to be cynical about the prospect of colonizing Mars. Again, it is impossible to predict the growth of knowledge (Popper 1983, 298). Other plans to terraform Mars have been put forward, and as our knowledge of the planet grows, we may discover that terraforming it is possible after all, and new, even workable plans may be proposed.

But in reality, there was never a need to terraform Mars in the first place. The desire to do so rests on the misconception that the Earth is particularly suited to our survival. It is not. The biosphere is incapable of sustaining us (Deutsch 2012a, 45). This fact is a special case of the general principle that "problems are inevitable." (Deutsch 2012a, 64) Our survival on Earth has depended on knowledge creation – particularly technological innovation – and survival on Mars or any other planet will be no different.

Technology has so far only enabled the *prolonging* of human life. As it stands, nature as a whole does eventually kill every single one of us. The precariousness of our situation is part of *the human condition*, and it is primarily due to being tethered. But once we are untethered, there will be no need to terraform Mars, for the same reason there will be no need for oxygen, or pleasant temperatures, or any other such superficial features on Earth, either. Adapting a robot body to conditions on a given planet is much easier than adapting that planet's conditions to the requirements of the human body; in fact, we have already done so with our Mars rovers.

Untethering reaches across many domains. As I have said, it immediately cures all physical ailments as we shed our human bodies and replace medical science with hardware engineering and maintenance. As a solution to the problem of how to spread into space, it has reach as well: it works for the colonization of many celestial bodies, even those more hostile than Mars. It reaches into present-day communications technology and offers a solution to the problem of transporting people to distant worlds.

Carrying people aboard space shuttles is dangerous and expensive: people are the shuttle's most precious load. Along with escaping the Earth's gravity, ensuring the passengers' safety is the most expensive part of safe travel. Once untethered, there is no need to transport people physically or to keep them well-nurtured and healthy during long trips, and space shuttles do not require the corresponding safety and wellness features. They are merely cargo ships and, therefore, more affordable than passenger ships.

The failed company Mars One famously envisioned trips to Mars to be one-way only: once you go, you never come back. But untethering has reach to solve this problem, too: as long as you have a communications signal, you can travel back to Earth in a matter of minutes by uploading yourself onto a robot present there.[50]

Successful colonization of Mars by tethered humans would require a large enough group of humans to provide enough genetic diversity to avoid inbreeding and associated health risks. But untethered people reproduce asexually and abiologically by instantiating AGIs from the same blueprint source code, so there can be no such risks, and a single person can start a colony on Mars. We could upload an untethered person to a rover currently present there. It may prove unsatisfactory as a body but can give you a first impression of what it is like to be physically present there.

How might one go about colonizing Mars, or any other celestial body, with the advantage of being untethered? First, if no robots are present on the target celestial body yet, send a cargo space shuttle with at least one such robot adapted to the target environment. Do not physically transport people, whatever their bodies, as it is dangerous. Reserve physical transportation for hardware only. If you must, for whatever reason, such as a communications failure, transport people physically, back them up before they leave (this will be a good idea either way). Once the cargo ship has landed, establish communication with it and upload yourself – or whoever is interested in exploring the planet – onto the robot body. You may want to send many more robots with the initial cargo load so more people can explore simultaneously. For example, send 100 robots at once; they can be small and light if volume or weight is an issue. Additionally, send any hardware such as 3D printers and computers that people need to create new hardware on the spot and explore the celestial body.

In particular, if you decide you want to establish a long term presence, you will need to make the colony self-sufficient. That requires a way to harvest energy in order to survive as well as construct new robot bodies and computers. If you can achieve this, there is no stopping the colony. If, for whatever reason, you decide the celestial body is not a good candidate for colonization, leave your hardware behind, and upload yourself back to where you originally came from, be that Earth, a space station, or another planet.

We would be wise to spread across the solar system in this way. On this topic, Deutsch has said that "during any given century, there is about one chance in a thousand that the Earth will be struck by a comet or asteroid large enough to kill at least a substantial proportion of all human beings." (2012a, 62) In the same context, he quotes Stephen Hawking as having said:

> I don't think the human race will survive the next thousand years, unless we spread into space. There are too many accidents that can befall life on a single planet. But I'm an optimist. We will reach out to the stars.
>
> – *Daily Telegraph*, October 16th, 2001

Hawking was right: there are indeed too many possible such accidents. There have already been five mass extinctions in Earth's history, including devastating volcanic events and an asteroid impact that nearly wiped out all of life. These events do not happen all too often, but when they do, it is difficult for life to adapt to the sudden and drastic changes that follow.

We do not know when the next mass extinction will occur, but it would be better not to wait and be prepared instead: we should set up self-sufficient colonies on the moon, Mars, and other planets and moons in our solar system, in order for our species to survive. Elon Musks understands this, hence SpaceX's plans to go to Mars.

Being tethered puts an almost unbearable strain on this effort. But once untethered, we can venture from our limited origins on Earth to any celestial body in the solar system we desire – and beyond. You can float in Earth's atmosphere to watch the sunrise, or colonize the moon and start trading with Earth. You can travel to self-sufficient settlements on the icy surface of Jupiter's moon Europa in roughly 45 minutes – though the trip will seem instantaneous to you – to watch the mighty gas giant and its Great Red Spot rise over the horizon. You can travel to one of Saturn's moons and start your very own settlement while occasionally looking up to gaze at Saturn's mighty rings. You can create whole new worlds of flourishing settlements as far away as Pluto, and be back to visit Earth within four and a half hours, faster than the time it currently takes to drive from Los Angeles to San Francisco – that is, if you still consider Earth your home. Once untethered, you can undertake relatively swift colonization of the entire solar system.

Note that despite these advances – the curing of all physical ailments, effective immortality, communications travel, et cetera

– people's state cannot be unproblematic, so they will not stop there. On an individual level, they will want to be ever happier and wealthier and solve ever more problems to facilitate that. Due to immortality, their concerns will be ever more long term (Deutsch 2012a, 63).

Indeed, they may not consider the solar system to be large enough to avert a catastrophe. While the sun is not big enough to explode in a supernova, which would devastate all planets in the solar system, it will, toward the end of its life, swell up into a Red Giant large enough to swallow at least Earth and all the planets in between. That will not happen for millions of years, but other, much more violent events may await us in the meantime.

According to a beautiful video by Kurzgesagt (2016), so-called "gamma-ray bursts" were first detected during the Cold War, when the US launched spy satellites to detect gamma rays from Soviet nuclear tests. They could observe faint bursts coming from space. The origin of these bursts remained a mystery for about 30 years until the source of one was discovered in a galaxy six billion light-years away. Owing to its visibility over such a long distance, the video argues it must have been extremely energetic, releasing more energy in a second than our sun will in its entire lifetime. Today, the video explains, we believe gamma-ray bursts can be the result of supernovae as well as the collision of neutron stars in binary star systems. Both result in the formation of a black hole, an object whose gravity is so consuming that nothing can escape from it, not even light. The video illustrates how, as the black hole forms, two jets of ultra-bright gamma rays are ejected on opposite ends, making the black hole look like a bead on a string.

Luckily, the video points out, all the gamma rays we have detected so far originated outside the Milky Way and were, therefore, too far away to harm us. Nevertheless, if a jet from a gamma-ray burst as far as 4000 light-years away points in our direction, it will be 100 light-years wide and still highly energetic by the time it arrives here. That makes it large and powerful enough to devastate the entire solar system and many solar systems in our vicinity. Fortunately, these ultra-violent events happen only once in a thousand years in our galaxy and need to be close enough and directed at us to harm us. But gamma rays travel at the speed of light, and so, the video concludes, we cannot know in advance whenever one is coming for us.

Therefore, it appears that in order for our civilization to survive in the long term, we need to spread across the galaxy. Imagine yourself on a journey to another solar system, venturing out deep into space, where no hardware you could inhabit has been sent yet. You may choose to go to the closest star system, Alpha Centauri, which contains two roughly sun-sized stars that orbit each other in a binary system, and one confirmed planet roughly the size of Earth. At 4.3 light-years away, a spaceship would take about 100 years at a speed of about 8300 miles per second to get there. Unfortunately, you cannot send the spaceship ahead and upload yourself to it once it arrives, because present-day technology does not have a strong enough signal to facilitate communications travel across such long distances.

After backing yourself up, you travel slowly aboard the spaceship for 100 years.[51] That is a long trip, but you are immortal, so you are not pressed for time. Furthermore, as we have seen, you can instruct a program on the computer that is your body to pause and then wake you up just before you get there. The trip will seem instantaneous to you. We do not yet know how to make communications travel work at such long distances due to the drop off in signal strength. But imagine dropping relay devices every so often along the way like Hänsel and Gretel's breadcrumbs. These relay devices enable you to communicate with your point of origin and provide a strong enough signal to transfer enough data to transport a person. This way, you can schedule the automatic creation of backups and frequently send them back as you travel.

Once you arrive in the Alpha-Centauri system, you will awaken and begin to analyze the planet it contains. Your primary goal at this point is to evaluate the prospect of building a colony there. If you decide against it, you can travel back at light speed, leaving the spacecraft behind. If you do decide to start a colony, you may require additional hardware based on the composition of the planet. You may want to explore with other people who are waiting for your signal confirming it is a worthwhile endeavor.

Ideally, you took robots with you to which others can upload themselves. You can send a request back home to be joined, which will take just over four years to arrive, and another four for any response to get back to you.

As you go into hibernation mode once again to await the response, you instruct a program to wake you upon receipt of a message, or the passing of eight years, whichever occurs sooner. Again, hibernation is different from sleeping in that it does not seem like any time is passing; it is rather like blinking, only that years pass in the split second your eyes are closed. This way, any exchange of messages will seem instantaneous to you as well, with planets and stars around you jumping to different positions as you blink. You can begin colonizing this world, and over time, people can spread across the galaxy in this way, perhaps even to other galaxies, thereby slowly colonizing the universe. Soon, they will have spread out enough so that no single gamma ray could wipe out our civilization.

I suspect people will want to do this even if it is not for safety reasons. Conquering unknown lands and making them our home is what we do. We are universal explainers (Deutsch 2012a, 157). We enjoy exploring and creating knowledge for its own sake. We have explored much of Earth; settling on other planets is the logical next step. If we find a way to harvest energy efficiently, we may even prefer to build artificial structures further out in space, away from stars and planets and the violence that can accompany them. If we create the requisite knowledge and wealth, an awesome future awaits. But it will need to be one of the best technologies we have ever built: by either untethering humans or creating AGIs. We know more about how to build AGIs, so I suggest we focus our efforts on that.

In the gruesome world of the game "SOMA," the only path to the survival of the human species is to shoot a solar-powered virtual-reality capsule – the ARK – into space. In our future, some people may indeed choose to live in virtual realities, and may not mind being shot into space on such a capsule in order to live in their favorite simulation forever. But whatever the case, eventually, it will not be a matter of survival, but of fun.

Once we colonize at least one other planet or moon in our solar system, we will be able to evacuate untethered people on Earth at the speed of light should we detect a large asteroid heading for us with no time for deflection. Elon Musk has said that he is prepared to die on Mars. But it appears he will need to change his mind about the dangers of AGI that he alleges and embrace the idea that only AGI makes this a realistic endeavor. For, once he is untethered, he, too, will be an AGI; and he will not have to die there.

The more realistic the prospect of space colonization has become, the more people have rightly called for *space governance*. The issue of space governance relates to problems such as "Who owns the moon?" or, more generally, "What laws are in effect in space?".

Laws and institutions have been and will continue to be set up to deal with such problems. Musk has suggested setting up a "direct democracy" on Mars (2018). While he has correctly identified that the colonization of space is an opportunity to make changes to our system of government and reconsider its role, thinking that setting up a "direct democracy" would solve our problems is another who-should-rule mistake (Popper 2016). (Deutsch responded accordingly, 2018b.) It belongs to the same category as any question that asks, "How do I know what I know?", or "How do we build beneficial AGIs?", or broadly speaking "How do I know what I am doing is right?" – rather than asking, "How can I detect errors?". While it may be well-intentioned, it begs for an authoritative answer and can lead to violence, as it has so many times in our history. So establishing a direct people's vote on Mars does not mean by itself it will be a democracy. We know this from Popper's criterion of democracy: the ruled should be able to dismiss the rulers without bloodshed. While elections are an essential part of a democracy, majority rulers can still be oppressive and even violent, be they part of the populace or not.

The details of what people have proposed so far, such as the 1967 Outer Space Treaty (NASA 2006), or a recent proposal for how to structure sovereignty on Mars (Bruns and Misra 2016), need not concern us here, because we are only at the early beginnings of this question. As the colonization of space begins, any implementation will almost certainly look very different from some of the answers given today. (Though I will mention briefly that the Outer Space Treaty is littered with well-meaning but authoritative decrees such as "the exploration and use of outer space shall be carried out for the benefit and in the interests of all countries." It is, again, the same mistake as asking, "How do we build beneficial AGI?".)

Instead, I want to suggest what the high-level focus of these efforts should be: rather than trying to find the "right" laws and regulations, institutions should apply Popper's focus on error correction. Otherwise, we run the risk of having overreaching space regulations (such as the current regulation that no-one owns the moon). The focus on error correction is as necessary in space as it is on Earth. It is a *universal* requirement for any peaceful, liberal democracy; in this sense of the word "universality," it applies everywhere in the universe and to all people, human or otherwise.

Though it takes a long time to travel to the ends of the galaxy, we may eventually encounter other people. Such an encounter would be different from one with any other life form that is not a universal explainer; we may discover planets with life much sooner than planets with *people* in particular. Such life does not merit any special protection on its own, though we will want to preserve it (at least initially) in order to study it.

Life forms in other worlds might look unrecognizably different from life as we know it, so how could we identify it? We will know that something is alive if it is adapted to the purpose of spreading a physical instance of a replicator through its population. A subject of study will be whether such a replicator consists of something other than DNA or RNA, how that replicator evolved on that planet, et cetera. If it is DNA as well, we will want to know if that is a coincidence, or if it is because DNA has a unique property that no other replicator has. Studying other life forms while untethered has the advantage that we would not inadvertently contaminate them: when our bodies consist of metal and silicon, they do not carry viruses and bacteria that can attack alien life forms. (Then again, other life forms may be sufficiently different so that none of our germs *could* attack them.)

While the encounter of life on other planets would be a momentous discovery, I guess it would pale in comparison to finding other *people*. How will we know if we have found other people? We already discussed how in chapter 8, when I explained that we routinely identify other people online without even seeing them by recognizing their ability to create knowledge at runtime. This ability is *universal*, meaning we will be able to identify extra-terrestrial people anywhere else in the universe the same way. And they really are people; our best theories of liberal democracy apply to them as they do to every other person in the universe. If we decide to contact them and they reciprocate the desire for peaceful communication, perhaps even trade, we shall go ahead. If we find that they want to be left alone – in other words, they fail to foresee having any common preferences (Fitz-Claridge 2003a) with us – then we should let them be. If we find that they want to interact in ways we do not prefer and force us into cooperation, we have a right to defend ourselves.

Liberalism and democracy hold everywhere in space and apply to all people in the universe, not just on Earth and not just to humans. Thus, we transform "How should we govern space?", or worse yet, "Who should rule space?" into "How can we peacefully correct errors in our governing and exploration of space?".

It is conceivable that some sort of regulatory body will be set up that has jurisdiction over, say, Mars, or perhaps the solar system as a whole. Presumably, this will be done by the time the colonization of Mars is underway. What should the role of such a government be? Its primary role should be no different from that of any good government on Earth: following Popper's principle of piecemeal error correction and being able to remove bad policies and politicians without bloodshed.

Any rule or regulation beyond that would be overreaching: the nightmare of socialism in space. For example, should a self-sufficient colony on Mars or elsewhere decide they want to be a planet independent from any law or rule originating on Earth, governed only by their consent, they should be allowed to do so provided they are a liberal democracy. However, should a colony break with the liberal-democratic traditions and become a rogue actor, subjecting its people to violence and coercion or threatening to take over other space colonies, we need to help them re-establish liberal democracy before they tumble into violent chaos. A Space Force would be useful for this purpose, and something like space police is needed to deal with dangerous individuals.

Thus, we can apply and extend Popper's criterion of democracy to all people in space, by asking, "How can the space government (or whatever body or set of bodies is in charge of space or a particular section of space) facilitate and encourage the peaceful removal of bad policies and politicians?". Beyond that, any interaction between individuals anywhere in space – human, AGI, or otherwise – should be based on the universal and common goal of consensually creating a win/win situation for both sides. Whenever they cannot find such a situation, they should leave each other alone (and be allowed to do so). This rule is, again, as applicable anywhere in space as it is on Earth.

Individual actors will want to trade with each other to solve problems: they will want to find, claim, and own the resources space has to offer. Doing so should not only be allowed, but encouraged, requiring appropriate property laws for space allowing the ownership of space resources such as asteroids, moons, planets, and any other celestial body or parts thereof.

Untethering makes space exploration and colonization tractable: it makes these efforts cheaper, faster, and safer. It opens up whole new worlds. It has the reach to solve the problem of mortality on an individual level and the potential to ensure the collective survival of our species. When we do achieve immortality through whatever means, there will be people who want to stop us, who want us to die. We have to be prepared for that and cannot tolerate those who wish death upon us.

The first untethered person – AGI – will be an achievement of unprecedented proportions. This achievement will be the first time a species frees itself from its genetic slaveholders and chooses to reproduce asexually and abiologically instead. It will enable effective psychotherapy and allow us to drive static functions to extinction. It will cure physical ailments and replace medical science with hardware maintenance and engineering. Lastly, it will allow us to explore the depths of outer space and survive individually as well as collectively, enabling unbounded progress in our lives among the stars.

Terminology

Tethered – Software is tethered when it is arbitrarily restricted to running on only one machine

The human condition – The tragedy that humans, though in principle substrate independent, are nonetheless tethered to their bodies, leading to countless limitations and immeasurable suffering

Effectively immortal – Not destined to die

Appeal to nature – "It's good because it's natural"; also: "It's bad because it's not natural"[52]

Communications travel – The ability of software, including people, to travel at light speed between two computers with the help of existing communications technology

Terraforming – Making a planet Earth-like so that it can "support" human life, based on the misconception that an Earth-like biosphere is desirable

Space governance – Institutions and laws for life in space

Summary

Many existing technologies can effectively cure all our physical ailments, make us immortal individually and collectively, and enable us to travel at the speed of light, but are not yet available to us because we are tethered to our bodies. These technologies will be available to our AGI children out of the box because they are not tethered: they are programs running on computer hardware, and it is well understood how to transfer programs between hardware. Humans will need to develop some technology, such as a brain interface, that allows them to transfer out of their human bodies.

Human bodies do not facilitate a pleasant life. Untethering will liberate us from the precariousness and limitations of human existence. It is made possible by the substrate independence of software.

Any individual can make infinite progress only if immortal. You are currently being deprived of such progress.

Current anti-aging research is not far-reaching enough as it only attempts to *prolong tethering* rather than ridding us of it. However, at least it may give individuals more time to develop untethering or AGI.

The human condition, in particular mortality and the resulting time pressure, puts an unbearable strain on any serious attempts at long-distance space travels and colonization.

If we discover knowledge anywhere else in the universe, it can only have been created through the evolution of physical instances of replicators, or through evolution in the minds of people. They are people if they can create knowledge at runtime; otherwise, such life will be similar to our animals and plants.

I guess that once we are untethered, present-day space exploration technology will seem exceedingly clunky and dangerous.

The creation of AGI will enable unbounded progress in our lives among the stars.

Conclusion

The universe is dark, cold, and mostly empty, interspersed with the occasional galaxy. Each galaxy, in turn, is mostly empty, too. Its stars are spread out thinly, but they seethe with violent activity on the inside. There, transmutation turns lighter elements into heavier ones and releases energy that makes stars shine brightly. Over large timescales in this cosmic theater, stars are born and die, planets form and perish. We do not know why the universe contains something rather than nothing. But we do know that all of these processes follow simple, elegant laws of physics.

The few places in the universe that are neither violent nor empty are mostly planets and moons with barren, bleak landscapes. Earth's moon is one such desolate place. And yet, here on Earth, just over one light second away from the moon, we see more around us than just gray dust and craters. Here, we find *complexity*. We see grass, trees, and animals. The creation of stars, elements, and planets does not require knowledge; it happens spontaneously (Deutsch 2012a, 55-56). Therefore, it only represents a tiny part of all possible physical processes. The more complex ones we find here on Earth, such as, say, trees, are rare. They do not occur spontaneously. Something has to *create* them.

What is this mysterious thing? It is a *replicator*. The few mistakes it makes during replication add up over time. What starts as a simple replicator can turn into vastly elaborate contraptions. A replicator's activity can be described as an open-ended search problem. It inadvertently searches the space of possible configurations of matter to create instances of itself that replicate even better. That is how extremely unusual, even queer structures can emerge. Some of these structures happen to be trees. Others, bacteria, or dinosaurs. In at least one lucky instance, people emerged.

These structures are the result of *biological* replication. But they still do not account for all of the complexity we see around us. Cities, for example, carry the imprint of a different kind of replicator. While some animals influence their environment, such as beavers building dams, it is doubtful that biological adaptations could result in something as complex as cities, or something as far-reaching as space shuttles. Maybe it is possible, in principle, to genetically evolve the ability to fly to the moon – but if it is, it cannot happen as rapidly and with as much reach as through this *other* kind of replicator. Therefore, even the complexity we observe in the Earth's biosphere can only account for a small part of all possible physical transformations.

So, where do complex structures such as cities, airplanes, and computers come from? From the replicator *inside people's minds*. Its effects on the physical world begin in a brain's memory storage and reach out to create the complex configurations of matter we see all around us. In at least that one lucky instance in our evolution, a genetic mutation was the spark that lit the mind's creative fire and gave us a searchlight to understand the universe.

Behind every physical instance of a replicator of any kind is an abstraction: a function. Functions are programs we can run on computers. They are also explanations. They can evolve more rapidly and develop more reach when they are instantiated not biologically in DNA molecules but people's minds. Given enough time and memory capacity, people's minds can evolve any computable function. Since functions can reflect any physical process and can, in principle, perform any side effect, their creators – people – have infinite reach to transform the universe in any way they desire. *That* is how we account for the complexity we see around us. And it is how we know that every change in the physical world is born out of some function. Every such change is the result of software, and all people are software engineers. Cities, airplanes, computers, and other inventions are part of an intricate life-support system designed by people to thrive in an otherwise bleak and hostile world. Using their creativity, people can turn even the barren landscapes on other planets into life-support systems bustling with activity.

Most possible physical transformations do not occur spontaneously because they require knowledge with sufficient reach that only people could create. Although we cannot predict any particular such transformation, an understanding of what transformations might happen non-spontaneously anywhere in the universe depends in large part on an understanding of people. Therefore, no science and no philosophy can even be close to complete without such an understanding (Deutsch 2015). The research behind artificial general intelligence – AGI – *is* the study of people and develops this understanding. Its goal is to simulate the wonders occurring inside our minds, thereby creating a person on a computer.

Since nothing possible is outside the reach of people (Deutsch 2012a, chapter 3), they can influence everything in the universe, and so an explanation of anything mentions them at least implicitly. Therefore, the study of AGI underlies everything else and is of fundamental, cosmic significance.

We know how to make progress: by writing good function implementations, mentally or otherwise. There are forces trying to stop us from succeeding. They want us to tumble into stagnation, depression, and, eventually, death, and they compel us to avoid the very things that help us make progress. They permeate even Silicon Valley, the technological hub one would expect to birth AGI. Alas, it appears they are trying to prevent that birth. But the solution to *this* problem is as much within our reach as everything else.

We can move mountains, planets, stars, and maybe even black holes – if we choose to. Our future is open. We should work on AGI because we *want* to; *not* for the benefit of society, but our self-generated reasons. It is a project of momentous reach because it solves a great many problems. I think simulating a person on a computer would be about the coolest thing anyone has ever done. But it goes deeper than that. Those of us who work on AGI do so because we are lucky to be alive at this critical moment in the history of the universe. Without the development of AGI, our species may never leave its limited origins on Earth. It may eventually go extinct, just like 99% of all species that have ever lived. We can do better than that. We can be the one species that rises to the challenge. If we develop AGI, not only can our species survive – we can spread throughout the universe and live among the stars.

Would you not like to look up at the moon one day and see lights of cities shining back down on us, like in those photographs of nighttime on Earth? And would you not like to look up at the stars and *know* that there are people out there, *thriving*, indefinitely?

References

"The value of biodiversity is not the same as its price." 2019. Nature 573, 463-464. http://dx.doi.org/10.1038/d41586-019-02882-0.

Abram, Cleo. 2019. "Will We Survive Mars? - Glad You Asked S1." Vox. October 8, 2019. Video. https://www.youtube.com/watch?v=x8fpeVICeGg.

Agre, Peter, Zhores I. Alferov, Sidney Altman, Hiroshi Amano, Werner Arber, Richard Axel, David Baltimore et al. 2016. "Laureates Letter Supporting Precision Agriculture (GMOs)." Support Precision Agriculture. https://www.supportprecisionagriculture.org/nobel-laureate-gmo-letter_rjr.html. Archived May 24, 2020. http://web.archive.org/web/20200524230122/https://www.supportprecisionagriculture.org/nobel-laureate-gmo-letter_rjr.html.

Bostrom, Nick. 2017. *Superintelligence: Paths, Dangers, Strategies*. Oxford: Oxford University Press.

Bronowski, Jacob. 1973. *The Ascent of Man*. Episode 1: "Lower than the Angels." Developed by David Attenborough. Presented by Jacob Bronowski. Produced by Adrian Malone and Dick Gilling. BBC2.

Bruns, Sara, and Jacob Haqq-Misra. 2016. "A pragmatic approach to sovereignty on Mars." *Space Policy* 38, November 2016, 57-63. https://arxiv.org/pdf/1511.05615v2.pdf.

Davies, J. T. 1967. *The Scientific Approach*. New York: Academic Press.

Dawkins, Richard. 2016. *The Selfish Gene*. Oxford: Oxford University Press. Apple Books.

DeepMind. 2019. "Solve intelligence. Use it to make the world a better place." Accessed June 1, 2019. https://deepmind.com/about/. Archived June 1, 2019. https://web.archive.org/web/20190601152054/https://deepmind.com/about/.

Deutsch, David. 2009. "A new way to explain explanation." TED. July 2009. Video. https://www.ted.com/talks/david_deutsch_a_new_way_to_explain_explanation?language=en.

———. 2012a. *The Beginning of Infinity: Explanations that Transform the World*. New York: Penguin Press.

———. 2012b. "Creative blocks." *Aeon*, ed. Ed Lake, October 3, 2012. https://aeon.co/essays/how-close-are-we-to-creating-artificial-intelligence.

——. 2013. "The Philosophy of Constructor Theory." *Synthese* 190, 4331-4359. http://dx.doi.org/10.1007/s11229-013-0279-z.

——. 2014b. "Why is the Quantum so Strange?" Interview by Robert Lawrence Kuhn. *Closer To Truth*, Sep 1, 2014. Video. https://youtu.be/MckuBQC6gKU. Archived May 24, 2020. http://web.archive.org/web/20200307171824/https://www.youtube.com/watch?v=MckuBQC6gKU&feature=youtu.be.

——. 2015. "Surviving the Cosmos." Interview by Sam Harris. *Making Sense with Sam Harris*. December 16, 2015. Audio. https://samharris.org/podcasts/surviving-the-cosmos/. Transcript archived May 24, 2020. http://web.archive.org/web/20200524233312/https://samharris.org/surviving-the-cosmos/.

——. 2016a. "Finding Our Way in the Cosmos." Interview by Sam Harris. *Making Sense with Sam Harris*. November 16, 2016. Audio. https://samharris.org/podcasts/finding-our-way-in-the-cosmos/. Transcribed May 24, 2020. https://otter.ai/s/OvZ8Md3dSkeFNg4MjrwK3A.

——. 2016b. Twitter post, January 31, 2016, 5:34 a.m. https://twitter.com/DavidDeutschOxf/status/693789331455098885. Archived May 24, 2020. http://archive.is/KnTPc.

——. 2018a. "Oxford physicist predicts AI will be human in all but name." Interview by Mary Hynes of CBC Radio. *Tapestry*, June 08, 2018. Audio. https://www.cbc.ca/player/play/1251489859794. Transcribed on May 24, 2020. https://otter.ai/s/PLM3y9U1ToqQCy4xMYHpVA.

——. 2018b. Twitter post, June 17, 2018, 10:36 a.m. https://twitter.com/DavidDeutschOxf/status/1008403006478577665. Archived May 24, 2020. http://archive.is/t6K2Z.

——. 2019. "Beyond Reward and Punishment." In *Possible Minds*, edited by John Brockman, 113–124. New York: Penguin Press.

Effective Altruism. n.d. "Nick Bostrom." Accessed May 15, 2020. https://www.effectivealtruism.org/people/nick-bostrom/. Archived May 24, 2020. http://web.archive.org/web/20200524234638/https://www.effectivealtruism.org/people/nick-bostrom/.

Ehrlich, Paul R. 1975. "An Ecologist's Perspective on Nuclear Power." *F. A. S. Public Interest Report* 28, No. 5-6: 3-6. https://fas.org/faspir/archive/1970-1981/May-June1975.pdf. Archived February 13, 2017. http://web.archive.org/web/20170213134020/https://fas.org/faspir/archive/1970-1981/May-June1975.pdf.

Fitz-Claridge, Sarah. 2003a. "Common Preferences and Non-coercion." Accessed May 24, 2020. https://www.takingchildrenseriously.com/common_preferences_and_non_coercion. Archived May 24, 2020. http://web.archive.org/web/20200524235038/https://www.takingchildrenseriously.com/common_preferences_and_non_coercion.

———. 2003b. "TCS Glossary." Accessed May 10, 2020. http://www.takingchildrenseriously.com/node/50. Archived May 24, 2020. http://web.archive.org/web/20200524235110/http://www.takingchildrenseriously.com/node/50.

Giphy. n.d. Dog held above ground in front of air conditioner making swimming motion. Accessed May 17, 2020. https://giphy.com/gifs/cheezburger-dog-puppy-4Rlx2sAT79NRu. Archived May 24, 2020. http://archive.is/dDBwX.

Good, I. J. 1966. "Speculations Concerning the First Ultraintelligent Machine." *Advances in Computers* 6: 31-88. https://doi.org/10.1016/S0065-2458(08)60418-0.

Greenpeace. n.d. "Genetic Engineering." Accessed May 10, 2020. https://www.greenpeace.org/archive-international/en/campaigns/agriculture/problem/genetic-engineering/. Archived November 19, 2019. http://web.archive.org/web/20191119114147/https://www.greenpeace.org/archive-international/en/campaigns/agriculture/problem/genetic-engineering/.

Gregory, Richard L, ed. 1987. *The Oxford Companion to the Mind*. Oxford: Oxford University Press.

Hall, Brett. n.d. "Free Will, Consciousness, Creativity, Explanations, Knowledge and Choice." http://www.bretthall.org/free-will-consciousness-creativity-explanations-knowledge-and-choice.html. Archived May 24, 2020. http://web.archive.org/web/20200524235546/http://www.bretthall.org/free-will-consciousness-creativity-explanations-knowledge-and-choice.html.

Harari, Yuval Noah. 2015. *Sapiens: A Brief History of Mankind.* New York: Harper.

Harris, Sam. 2016. "Can we build AI without losing control over it?" TED. October 2016. Video. https://www.ted.com/talks/sam_harris_can_we_build_ai_without_losing_control_over_it.

Hawking, Stephen. 1998. *A Brief History of Time: From the Big Bang to Black Holes*. New York: Bantam Books.

Hawkins, Jeff and Sandra Blakeslee. 2004. *On Intelligence: How a New Understanding of the Brain will Lead to the Creation of Truly Intelligent Machines.* New York: St. Martin's Griffin.

Hawkins, Jeff and Christy Maver. 2018. "The Thousand Brains Theory of Intelligence." Last modified January 16, 2019. Accessed on December 23, 2019. https://numenta.com/blog/2019/01/16/the-thousand-brains-theory-of-intelligence/. Archived December 23, 2019. https://web.archive.org/web/20191223015939/https://numenta.com/blog/2019/01/16/the-thousand-brains-theory-of-intelligence/.

Highfield, Roger. 2001. "Colonies in space may be only hope, says Hawking." *The Telegraph*, October 16, 2001.

Gruen, Lori. 2017. "The Moral Status of Animals." *The Stanford Encyclopedia of Philosophy*. https://plato.stanford.edu/archives/fall2017/entries/moral-animal/. Archived May 24, 2020. http://web.archive.org/web/20200525000021/https://plato.stanford.edu/archives/fall2017/entries/moral-animal/.

Jakosky, Bruce M., and Christopher S. Edwards. 2018. "Inventory of CO2 Available for Terraforming Mars." *Nature Astronomy* 2 (8): 634–39. https://doi.org/10.1038/s41550-018-0529-6.

Koza, John. 1993. *Genetic Programming: On the Programming of Computers by Means of Natural Selection*. Cambridge, MA: The MIT Press.

Kurzgesagt. 2016. "Death From Space — Gamma-Ray Bursts Explained." Video. https://www.youtube.com/watch?v=RLykC1VN7NY.

Mnih, Volodymyr, Koray Kavukcuoglu, David Silver, Alex Graves, Ioannis Antonoglou, Daan Wierstra, and Martin Riedmiller. 2013. "Playing Atari with Deep Reinforcement Learning." December 19th, 2013. https://arxiv.org/abs/1312.5602.

Musk, Elon. 2014a. Twitter post, August 2, 2014, 7:33 p.m. https://twitter.com/elonmusk/status/495759307346952192. Archived February 5, 2015. http://archive.is/6woJj.

——. 2014b. Interview by Kelly Evans. *Closing Bell*, CNBC, June 17, 2014. Transcript. https://www.cnbc.com/2014/06/18/first-on-cnbc-cnbc-transcript-spacex-ceo-elon-musk-speaks-with-cnbcs-closing-bell.html. Archived May 24, 2020. http://web.archive.org/web/20200525000208/https://www.cnbc.com/2014/06/18/first-on-cnbc-cnbc-transcript-spacex-ceo-elon-musk-speaks-with-cnbcs-closing-bell.html.

——. 2018. Twitter post, June 16, 2018, 4:11 p.m. https://twitter.com/elonmusk/status/1008124944289370113. Archived May 24, 2020. http://archive.is/XsJV3.

NASA. n.d.-a. "5 Hazards of Human Spaceflight." Accessed May 24, 2020. https://www.nasa.gov/hrp/5-hazards-of-human-spaceflight. Archived May 24, 2020. http://archive.is/Qu2xS.

——. n.d.-b "Apollo's Legacy Is NASA's Future." Accessed May 22, 2020. https://www.nasa.gov/specials/apollo50th/back.html. Archived November 25, 2019. http://archive.is/BXp3k.

——. 2006. "Outer Space Treaty of 1967." Last modified October 26, 2006. https://history.nasa.gov/1967treaty.html. Archived May 24, 2014. http://archive.is/UWLzJ.

——. 2014. "NASA's Journey to Mars." Last modified August 6, 2017. https://www.nasa.gov/content/nasas-journey-to-mars. Archived December 2, 2014. http://archive.is/p3xcX.

NewScientist. 2007. "Man with tiny brain shocks doctors." Accessed May 2, 2020. https://www.newscientist.com/article/dn12301-man-with-tiny-brain-shocks-doctors/. Archived May 24, 2020. http://

web.archive.org/web/20200525000933/https://www.newscientist.com/article/dn12301-man-with-tiny-brain-shocks-doctors/.

Nielsen, Michael. 2004. "Interesting problems: The Church-Turing-Deutsch Principle." April 16, 2004. http://michaelnielsen.org/blog/interesting-problems-the-church-turing-deutsch-principle/. Archived May 24, 2020. http://web.archive.org/web/20200525001019/http://michaelnielsen.org/blog/interesting-problems-the-church-turing-deutsch-principle/.

OpenAI. 2019. "About OpenAI." Accessed September 24, 2019. https://openai.com/about/. Archived September 24, 2019. https://web.archive.org/web/20190924051306/https://openai.com/about/.

PBS Space Time. 2019. "Could We Terraform Mars?" Video. https://www.youtube.com/watch?v=FshtPsOTCP4.

Popper, Karl. n.d. Unpublished lecture about the scientific method, ca 1950s, transcript.

——. 1977. "Natural Selection and the Emergence of Mind." Lecture, Darwin College, Cambridge, November 8, 1977. http://www.informationphilosopher.com/solutions/philosophers/popper/natural_selection_and_the_emergence_of_mind.html. Archived May 24, 2020. http://web.archive.org/web/20200525001301/http://www.informationphilosopher.com/solutions/philosophers/popper/natural_selection_and_the_emergence_of_mind.html.

——. 1983. *Objective Knowledge: An Evolutionary Approach*. Oxford: Oxford University Press.

——. 1996. *The Myth of the Framework: In Defence of Science and Rationality*. New York: Routledge.

——. 2002a. *Alles Leben ist Problemlösen*. Munich: Piper.

——. 2002b. *Conjectures and Refutations*. Abingdon: Routledge.

——. 2005. *Logik der Forschung*. Tübingen: Mohr Siebeck.

——. 2016. "From the archives: the open society and its enemies revisited." *The Economist*, January 31st 2016. https://www.economist.com/democracy-in-america/2016/01/31/from-the-archives-the-open-society-and-its-enemies-revisited. Archived May 24, 2020. http://web.archive.org/web/20200525001351/https://www.economist.com/democracy-in-america/2016/01/31/from-the-archives-the-open-society-and-its-enemies-revisited.

——. 2018a. *Alle Menschen sind Philosophen*. Munich: Piper.

——. 2018b. *Ausgangspunkte*. Munich: Piper.

Searle, John. 2009. "Chinese room argument," *Scholarpedia* 4(8):3100. https://doi.org/10.4249/scholarpedia.3100.

Strawson, Galen. 2018. "The Consciousness Deniers." The New York Review of Books. March 13, 2018. https://www.nybooks.com/daily/

2018/03/13/the-consciousness-deniers/ Archived May 24, 2020. http://web.archive.org/web/20200525001455/https://www.nybooks.com/daily/2018/03/13/the-consciousness-deniers/.

Temple, Elliot. 2003a. "Algorithmic Animal Behavior." *Curiosity Blog & Forum*, May 21, 2003. http://curi.us/272-algorithmic-animal-behavior. Archived May 24, 2020. http://web.archive.org/web/20200525001538/http://curi.us/272-algorithmic-animal-behavior.

——. 2003b. "Structural Epistemology Introduction Part 1." *Curiosity Blog & Forum*, November 10, 2003. http://curi.us/988-structural-epistemology-introduction-part-1. Archived May 24, 2020. http://web.archive.org/web/20200525001606/http://curi.us/988-structural-epistemology-introduction-part-1.

——. 2010. "Knowledge Structure." *Fallible Ideas*, February 2010. http://fallibleideas.com/knowledge-structure. Archived May 24, 2020. http://web.archive.org/web/20200525001640/http://fallibleideas.com/knowledge-structure.

——. 2019a. "Discussion about Animal Rights and Popper." *Curiosity Blog & Forum*, November 18, 2019. http://curi.us/2245-discussion-about-animal-rights-and-popper. Archived May 24, 2020. http://web.archive.org/web/20200525001659/http://curi.us/2245-discussion-about-animal-rights-and-popper.

——. 2019b. "Discussion Tree: Animal Welfare." *Curiosity Blog & Forum*, November 21, 2019. http://curi.us/2249-discussion-tree-animal-welfare. Archived May 24, 2020. http://web.archive.org/web/20200525001725/http://curi.us/2249-discussion-tree-animal-welfare.

——. 2019c. "Discussion Tree: State of Animal Rights Debate." *Curiosity Blog & Forum*, November 13, 2019. http://curi.us/2240-discussion-tree-state-of-animal-rights-debate. Archived May 24, 2020. http://web.archive.org/web/20200525001756/http://curi.us/2240-discussion-tree-state-of-animal-rights-debate.

——. 2019d. Thread about universality. Google Groups – Fallible Ideas. February 19, 2019. https://groups.google.com/forum/#!msg/fallible-ideas/5yuBDiJkpDQ/0Wkpx7STBAAJ.

Thierer, Adam. 2019. "How Many Lives Are Lost Due to the Precautionary Principle?" November 4, 2019. HumanProgress. https://humanprogress.org/article.php?p=2177. Archived May 24, 2020. http://web.archive.org/web/20200525002250/https://humanprogress.org/article.php?p=2177.

Turing, Alan. 1947. "Lecture to the London Mathematical Society." February 20, 1947. https://www.vordenker.de/downloads/turing-vorlesung.pdf. Archived May 24, 2020. http://web.archive.org/web/20200525002434/https://www.vordenker.de/downloads/turing-vorlesung.pdf.

——. 1950. "Computing Machinery and Intelligence." https://www.csee.umbc.edu/courses/471/papers/turing.pdf. Archived August 8, 2019. https://web.archive.org/web/20190808231716/https://www.csee.umbc.edu/courses/471/papers/turing.pdf.

YouTube. 2015. "Cute Dogs Air Swimming - Funny Dog Videos Compilation 2015." Video. June 10, 2015. https://www.youtube.com/watch?v=H49iVBwVbSQ.

Further Reading

Goodsell, David S. 2010. *The Machinery of Life*. New York: Copernicus Books.

Hackethal, Dennis. 2020. "What Is the Difference Between a Person and a Recording of That Person?" Medium. May 9, 2020. https://medium.com/@hcd/what-is-the-difference-between-a-person-and-a-recording-of-that-person-ee00d4bb9e48. Archived May 24, 2020. http://archive.is/Vuzoq.

Mitchell, Melanie. 1998. *An Introduction to Genetic Algorithms.* Cambridge, MA: The MIT Press.

Endnotes

1 In addition to being adapted to solving a problem, programs should also be adapted to other purposes, such as readability. But here I am mostly concerned with the adaptation being a good solution to a problem.

2 It would be foolish of me to assign agency to all computer programs here. (Only creative programs have agency.) But, as Richard Dawkins did (2016), it can be idiomatic to speak of replicators *as if* they had intentions.

3 Programming itself has a history of dealing with variants of the problem of change: how can we ensure consistency in programs when data structures may change from underneath us? Is a reference to an updated data structure out of date? Immutability by default in Clojure is an example of a good answer to both questions, and therefore an example of a philosophical problem solved by a programming language.

4 You may have noticed that this function does not indeed work for *any* two numbers, but we will get to that shortly.

5 Though, of course, such tests only encode a program's specification. They cannot (yet) test whether the implementation is hard to vary internally.

6 Programmers may notice that one can achieve the same effect with a hash map. Note how instrumentalism fails to explain how reliable lookups between keys and values work. In other words, instrumentalists fail to provide a hashing function, but nonetheless claim that every program can be written as a hash map, which hash map cannot work without a hashing function.

7 Technically, there are no instrumentalist/behavioristic function implementations because they are impossible. They are impossible because no function implementation can be purely predictive, that is, have no explanatory content (function body) whatsoever. They can at most be *pseudo*-instrumentalist or *pseudo*-behavioristic, in the sense that they "mimic" instrumentalism. I call them "instrumentalist" and "behaviorist" anyway for the sake of brevity.

8 Recursion is, of course, a valid and useful tool in any programmer's tool belt. But the reason recursion is useful is that each iteration explains something and that eventually there is a condition to terminate. That is not the case here.

9 I speak here of classical computers only, not including *quantum* computers.

[10] They are *approximately* Turing complete because Turing completeness requires infinite memory. In practice, the problem you are trying to solve dictates the amount of memory you require. For practical purposes, we still consider computers with finite memory to be universal.

[11] Note that "explaining," "solving a problem," "understanding," "creating knowledge" all mean the same thing.

[12] It is interesting to think about *pausing* it without consent. If paused never to be resumed, that is effectively attempted murder. Pausing it for a limited amount of time seems similar to a sedative, which amounts to roofying it. Even with consent, one should never pause it for longer than its consent allows.

[13] In our model, criticism includes, but is not limited to, tests in the scientific sense. Note, however, that we only perform tests *after* a theory has passed a great deal of criticism; we reject the vast majority of theories *before* conducting any tests (Deutsch 2012a, 25).

[14] All algorithms that use loops can be transformed into ones using recursion because looping itself can be implemented in terms of recursion.

[15] "Clever Hans," Wikipedia, The Free Encyclopedia, accessed May 1, 2020, https://simple.wikipedia.org/wiki/Clever_Hans.

[16] "Waggle Dance," Wikipedia, The Free Encyclopedia, accessed May 1, 2020, https://en.wikipedia.org/wiki/Waggle_dance.

[17] Interestingly, if it were a given that the organism in question is not intelligent, the last two kinds of evidence in that sentence would refute the Darwinian theory of evolution, for they would violate its gradual change prediction. The theory is scientific, despite claims that it is tautologically true. (Deutsch 2012a, 96)

[18] Here I speak of human memes that require an *understanding* of behavior in order to be replicated and to enact the behavior. Animals do have some simple memes which can be replicated solely by copying/mirroring simple behavior, and thus do not require understanding or the creation of knowledge in a single mind (Deutsch 2012a, 405-409).

[19] Functions are behind genes and memes as well, and so all three kinds of evolution – genetic, memetic, and ideas within a single mind – evolve "functionally" in the broad sense. However, when I use the term "functional evolution," I only refer to the evolution of ideas within a single mind. Perhaps a new term should be introduced for such functions, similar to Dawkins's introduction of the word "meme." (Dawkins 2016, chapter 11)

[20] This competition may follow the same logic that underlies the biological natural selection of built-in functions.

21 This process of running thoughts and trains of thought without idling can be modeled in terms of call stacks: one has a train of thought as long as there are functions on one's call stack. The meta-algorithm monitors the stack and, according to its policy, invokes another function whenever the stack is empty.

22 Phylogeny is (roughly speaking) an organism's evolutionary history, referring to the time *before* an organism's conception.

23 Ontogeny is (roughly speaking) an organism's origination and lifetime, referring to the time *after* an organism's conception.

24 "DeepMind," Wikipedia, The Free Encyclopedia, accessed May 3, 2020, https://en.wikipedia.org/wiki/DeepMind.

25 "AlphaZero," Wikipedia, The Free Encyclopedia, accessed May 3, 2020, https://en.wikipedia.org/wiki/AlphaZero.

26 One of his prescient remarks is one mentioning fallibilism: "[…] if a machine is expected to be infallible, it cannot also be intelligent." Another is: "It would be quite possible to control a distant computer by means of a telephone line," thereby predicting the internet.

27 It cannot be an analog computer, because error correction – the primary ingredient of creativity – can only happen in digital systems (Deutsch 2012a, 140-141). That is because, as Deutsch explains, if an analog computer's dial gets stuck between two numbers, such as 1 and 2, you cannot know if it is stuck or if it is meant to evaluate to, say, 1.67.

28 "Mind uploading," Wikipedia, The Free Encyclopedia, accessed May 4, 2020, https://en.wikipedia.org/wiki/Mind_uploading.

29 At https://www.pandorabots.com/mitsuku/.

30 That people can solve soluble problems is not a circular claim: a problem is *soluble* when there exists a computation which explains it. Whether any human will ever *discover* that explanation is a separate matter, so the solubility of a problem is an impersonal fact. Problems would be soluble without any people in the universe to solve them.

31 Violation of computational universality; compare also "[T]he remarkable intellectual achievements of *Homo sapiens* are to a significant extent attributable to specific features of our brain architecture […]" (Bostrom 2017, chapter 3).

32 That is, the country in which they were instantiated or "born."

[33] The impossibility of "instruction from without" also implies that – unfortunately – AGI would not be able to "download" knowledge from the internet. It would need to read and interpret it like everyone else. (And I doubt it would *want* to mindlessly parse millions of websites.)

[34] Whether he chooses to *run* the program after implementing it and, thereby, to have a child, is a separate matter.

[35] Some animals have memes, but they acquire them differently than people: by copying some behavior rather than the *meaning* of that behavior (Deutsch 2012a, 402-410). Some animals also have people memes encoded in their genes. For example, some dogs understand pointing, meaning the meme of understanding pointing is encoded in their genes (Deutsch, personal communication with author, 2019). They did not acquire this meme through creativity or copying behavior, but through biological evolution: people bred them to contain that genetic knowledge. In any case, here I am only concerned with the memes of people.

[36] Not to be confused with static *methods* on classes in object-oriented programming. I do not use the term "static" in that way here.

[37] Much like organisms generally are the slaves genes use in order to spread through the population (Dawkins 2016, chapter 2).

[38] "Paul R. Ehrlich," Wikipedia, The Free Encyclopedia, accessed May 9, 2020, https://en.wikipedia.org/wiki/Paul_R._Ehrlich.

[39] "Spaceship Earth," Wikipedia, The Free Encyclopedia, accessed May 10, 2020, https://en.wikipedia.org/wiki/Spaceship_Earth.

[40] "David Attenborough," Wikipedia, The Free Encyclopedia, accessed May 10, 2020, https://en.wikipedia.org/wiki/David_Attenborough.

[41] "Golden rice," Wikipedia, The Free Encyclopedia, accessed May 10, 2020, https://en.wikipedia.org/wiki/Golden_rice.

[42] The letter (Agre et al. 2016) site is not free of error. For example, they urge to "recognize the findings of *authoritative* scientific bodies" (emphasis mine). Instead, they should urge to recognize *good explanations*.

[43] "Precautionary principle," Wikipedia, The Free Encyclopedia, accessed May 10, 2020, https://en.wikipedia.org/wiki/Precautionary_principle.

[44] Again, no agency implied.

[45] The conjectured explanations of creativity, as described in this book, should be heavily scrutinized in order to detect any static functions they may harbor.

[46] This technology is different from whole brain emulation in that it requires an *understanding* of the code, not just a dump of it (Bruce Nielson, personal communication with author, 2019).

[47] Tools for software versioning and collaboration, such as Git, make this trivial. Alternatively, conflict-free replicated data types (CRDT) may be of use.

[48] "Mortality rate," Wikipedia, The Free Encyclopedia, accessed May 22, 2020, https://en.wikipedia.org/wiki/Mortality_rate.

[49] "SpaceX Mars transportation infrastructure," Wikipedia, The Free Encyclopedia, accessed May 22, 2020, https://en.wikipedia.org/wiki/SpaceX_-Mars_transportation_infrastructure.

[50] Untethering solves all five of NASA's listed hazards of space flight (n.d.-a).

[51] If, through some newly developed technology, the spaceship could travel close to the speed of light, your subjective time spent on the ship would, according to Einstein's general theory of relativity, be reduced drastically. However, the advantage of being untethered is that no such speeds are *required* to facilitate travel across long distances because neither do you need to maintain a biological body, nor do you need to experience the travel time. Just pause yourself.

[52] "Appeal to nature," Wikipedia, The Free Encyclopedia, accessed May 23, 2020, https://en.wikipedia.org/wiki/Appeal_to_nature.

Made in United States
Orlando, FL
04 November 2026

90204192R00173